An Introduction to Biotechnology

A Genetic Manipulation Perspective

An Introduction to Biotechnology

A Genetic Manipulation Perspective

Editor

RUP LAL

Professor

Department of Zoology

University of Delhi, Delhi

TECHSAR™

NEW DELHI

Published and Printed by
TechSar LLP
4435-36/7, Ansari Road, Daryaganj
New Delhi-110 002 (India)
E-mail: tech_info@techsarworld.com
Website: www.techsarworld.com

ISBN: 978-93-84588-86-1

Reprint 2026

Preface

The term Biotechnology was coined in the year 1919 by a Hungarian civil engineer Karl Ereky. At that time the term biotechnology meant the use of microorganisms for the production of biomolecules for the benefit of human beings. However, the progress in biotechnology remained slow up to 1970s as limited tools for manipulating DNA were available. The advancement and discoveries of recombinant DNA techniques especially development of cloning vectors, transformation methods, DNA sequencing and Polymerase Chain Reaction in the early 1980s created a strong base for the rapid growth of biotechnology. By using these tools several beneficial products like recombinant insulin, human growth hormone, several enzymes and other products of commercial importance were produced.

Subsequent developments and the benefits from biotechnology coupled with rapid advances in gene manipulation led to increased interest in the field of biotechnology. As a result these tools have now been introduced in the curricula of undergraduate and postgraduate students. However, there is no elementary book available that describes these techniques for the beginners. Keeping this in view we have compiled a short handbook which can act as a primer for the protocols in biotechnology and gene manipulations. This book was written by Ph.D. scholars while working in laboratory by incorporating their experiences in a way that makes it easy to understand and self-explanatory for the beginners. The book starts with an introduction to modern biotechnology, followed by introduction to gene manipulation. The subsequent chapters illustrate and describe the protocols for nucleic acid isolation, protein isolation, restriction digestion, polymerase chain reaction, primer designing, ligation, cloning, transformation, Southern hybridization and DNA sequencing.

We hope that both undergraduate and postgraduate students will find this book useful. The authors gratefully acknowledge all those who have been directly or indirectly associated with conception and execution of this idea and finally giving it a shape of a book.

Rup Lal

Contributors

- **Amit Kumar Singh** - amit.k.singhbhu@gmail.com
- **Anjali Saxena** - anjali.saxena1403@gmail.com
- **Ankita Dua** - ankitadua@gmail.com
- **Anukriti Sharma** - anukriti.sharma8@gmail.com
- **Charu Tripathi** - charutripathi89@gmail.com
- **Helianthous Verma** - helianthousverma@gmail.com
- **Mansi Verma** - mansiverma20@gmail.com
- **Namita Nayyar** - nayyarnamita@gmail.com
- **Nitish Kumar Mahato** - nitishyayati.du@gmail.com
- **Pooja Rani** - poojasrivastav@gmail.com
- **Princy Hira** - princy.hira@gmail.com
- **Priya Singh** - priya.impulse@gmail.com
- **Puneet Kohli** - puneet.k1311@gmail.com
- **Rashmi Kumari** - rash2189@gmail.com
- **Roshan Kumar** - roshanzhc@gmail.com
- **Shailly Anand** - shailly.anand@gmail.com
- **Udita Mukherjee** - udita297@gmail.com
- **Utkarsh Sood** - utkarsh.sood@gmail.com
- **Vipin Gupta** - guptavipin142@gmail.com
- **Vivek Negi** - viveknegi2006@gmail.com

Contents

Introduction to Biotechnology

Shailly Anand, Mansi Verma, Udita Mukherjee,
Princy Hira & Utkarsh Sood

The term 'biotechnology' was coined by a Hungarian engineer, Karl Ereky in 1919. At that time the term biotechnology meant the production of raw materials with the aid of living organisms especially microorganisms. Biotechnology is not new to humanity and man has been manipulating living things for his own concern. In the late eighteenth century and the beginning of nineteenth century, vaccination, crop rotation and selective animal breeding were practiced. However, the discovery of microorganisms, followed by Mendel's laws of genetics during the end of nineteenth century provided further impetus to biotechnology. Initial focus was on the use of microorganisms for the production of antibiotics and other products through fermentation processes. The progress in biotechnology was slow during the first half of the twentieth century. Even after the discovery of double helical structure of DNA by Watson and Crick in 1953 (Table 1), there was not much progress until 1970. The era of modern biotechnology began in early 1970s with the discovery of simpler tools for manipulating DNA. This was eventually followed by the discovery of a number of newer tools that virtually provided a strong momentum to the growth of biotechnology.

Significant Biotechnology Milestones

Table 1 Significant Milestones in the History of Biotechnology and Its Major Tools from 1953 Onwards

Year	*Milestones*
1953	- Discovery of the molecular structure of DNA by James Watson and Francis Crick.
1966	- The genetic code is "cracked" - showing that a sequence of three nucleotide bases determines each of 20 amino acids (two more amino acids have since been discovered).
1973	- Stanley Cohen and Herbert Boyer performed the first experiments in genetic engineering by developing a technique to clone segments of DNA molecules.
1976	- Genentech, the first company devoted to producing genetically engineered products, established in San Francisco, California.
1977	- Sanger's dideoxy method of DNA sequencing developed. - Maxam-Gilbert's method of DNA sequencing by chemical synthesis of DNA developed.

Contd...

Contd...

1979	- Genetic engineering used to synthesize human insulin.
1982	- First recombinant DNA vaccine for livestock developed. - First biotech drug approved by FDA: human insulin produced in genetically modified bacteria. - First genetic transformation of plant cell: Petunia.
1983	- Kary Mullis conceives Polymerase Chain Reaction (PCR), a method of making multiple copies of parts of DNA. This process has become one of the key tools in genetics.
1984	- DNA fingerprinting technique developed by Alec Jeffreys. - The entire genome of the Human Immunodeficiency Virus (HIV) cloned and sequenced. - Robert Sinsheimer proposes mapping of all the genes in the human genome. The proposal led to the initiation of the Human Genome Project.
1985	- Transgenic plants resistant to insects, viruses and bacteria were field-tested for the first time. - The National Institute of Health (NIH) approved guidelines for performing gene-therapy experiments in humans.
	- First recombinant vaccine developed: Hepatitis B. - First anti-cancer drug produced through gene manipulation: Interferon. - The U.S. government publishes the *Coordinated Framework for Regulation of Biotechnology*. - A University of California Berkeley scientist describes how to combine antibodies and enzymes to create pharmaceuticals.
1987	- First approval for field-test of modified food plants: virus-resistant tomatoes.
1988	- First U.S. patent awarded for a biotech animal - a transgenic mouse. - Congress funds the Human Genome Project (HGP) to support mapping the human genome.
1989	- First approval for field test of biotech cotton: insect-protected cotton. - Plant Genome Project begins. - The Human Genome Project is launched on January 3. James Watson is the first director.
1990	- The first GENE THERAPY experiment is performed on a four-year-old girl with adenosine deaminase deficiency. The procedure involves inserting genetically altered cells into the veins of the patient.
1995	- First bacterial GENOME, *Haemophilus influenzae,* was completely sequenced by the Institute for Genomic Research.
1998	- Human embryonic stem cell lines are established. - The first complete animal genome, for the *Caenorhabditis elegans* worm sequenced.

2000	- Rough draft of the human genome sequence announced. - First complete map of a plant genome (*Arabidopsis thaliana*) developed.
2001	- First complete map of the genome of *Oryza sativa* (rice) plant completed.
2002	- Researchers announce successful results for a vaccine against cervical cancer, the first demonstration of a preventative vaccine for a type of cancer. - International consortia sequence the genomes of the parasite that causes malaria and the species of mosquito that transmits the parasite. - FDA approves 20 products in 2002.
2003	- Human Genome Project completed. - FDA approves 25 products in 2003, many of which are "firsts" - first biologic approved for psoriasis, first fusion inhibitor approved for HIV/AIDS, first drug for mild to moderate Alzheimer's disease, first drug approved for Fabry's disease and first biologic approved for asthma. - Global acceptance of biotech crops continues to rise: biotech crops are grown on more than 167 million acres farms by 7 million farmers in 18 countries. - President George W. Bush speaks at the BIO 2003 Annual Convention applauding the efforts of the biotech industry for furthering biomedical, agricultural biotechnology, and biodefense efforts.
2004	- The map based sequence of rice genome published (International Rice Genome Sequencing Project. India's first ever contribution in genome project by: University of Delhi , South Campus and Indian Agricultural Research Institute , IARI, New Delhi).
2005	- CraigVenter popularized the metagenomic approach to study biodiversity.
2006	- Pyrosequencing developed/DNA sequencing.
2007	- Craig Venter Institute created the first *Synthetic Bacteria* and applied for patent. - Institute conducts complete genome swap changing one species of bacteria to another. - *First individual human genome* of Craig Venter sequenced.
2010	- First *synthetic genome* made by Craig Venter.
2012	- *Human Microbiome Project* completed.
2013	- *Artificial Kidney* created by IIT-Delhi alumni gets ready for laboratory trials.
2014	- *Cancer Spit Test* – A device designed by researchers at the University of California detects oral cancer from a single drop of saliva (no need of biopsies).

DISCOVERY OF NEW GENETIC TOOLS

In its new form, the term "biotechnology" refers to the use of living organisms or their products to modify human health and human environment. Biotechnology, in general terms, is also known as gene cloning, genetic engineering or molecular cloning. It is in fact an umbrella term encompassing a number of techniques. In other words, biotechnology primarily includes cloning, characterization and expression subsequent to manipulations of genes. All this became possible due to three primary tools, i.e., transformation methods (Fig. 1), cloning vectors and restriction endonucleases (Fig. 2). With the availability of basic tools in 1970s, the discipline of biological sciences witnessed a revolution. This technology has given a number of techniques that are now used directly or indirectly for tracing, isolating and characterizing genes. Although the discovery of these three tools brought a revolution in the field of biological sciences, the role of other techniques should not be underestimated!

Essence of Biotechnology: The Recombinant DNA Technology

Biotechnology relies on two core technologies, namely, genetic engineering and biochemical engineering. Genetic engineering, popularly known as recombinant DNA technology involves isolation, manipulation, combination, addition of DNA that results in the product of interest. This forms the essence of biotechnology. On the other hand, biochemical engineering requires large scale growth of beneficial microbes by tissue culture techniques or fermentation, so as to obtain the desired product in huge quantities.

Recombinant DNA technology therefore harbors on *in vitro* genetic manipulations by man. By this, the gene of interest is obtained from its source, is cut into smaller fragments and inserted into a cloning vehicle that helps to carry it into the host cell. Transformation into the host cell results in the production of the protein of interest (beneficial to human). Below is a step-by-step guide to create a recombinant DNA molecule.

1. **Isolation –** This is the first step in the process in which either the source genomic DNA can be isolated or a specific gene may be obtained by polymerase chain reaction and subsequent purification.
2. **Selection of a cloning vector –** There are diverse vectors available that differ in the amount of insert capacity. Therefore, based on the insert size, a suitable vector may be selected. In the subsequent section different types of cloning vectors have been described.
3. **Fragmentation –** It deals with cutting of the source DNA into smaller fragments. This can be done either by restriction

endonucleases (that cleave a specific sequence within the DNA) or by chemical means. This step also involves cutting of the vector by a suitable restriction enzyme.

4. **Ligation** – During this step, the vector and the smaller DNA fragments are joined together. This is mediated by DNA ligases that form a phosphodiester bond between the two. The cloning vector carrying the foreign gene is referred as the recombinant DNA (rDNA).
5. **Transformation** – This entails the introduction of the recombinant molecule generated above into a host cell. There are many methods available to accomplish this task. It involves the process of making the cell membrane of recipient cells transiently porous to allow the exogenous DNA to move in. Application of electric impulses (i.e., electroporation) or chemical treatment of cells ($CaCl_2$) followed by a heat shock, are the general methods of choice. Other methods include microinjection, particle bombardment, liposome mediated gene transfer, calcium phosphate precipitation, transposon mediated transfer, transduction, etc.
6. **Selection** – This is one of the most crucial steps in the rDNA technology which allows the selection of the cells with recombinant DNA from those untransformed. Here comes the use of selectable marker genes present on the cloning vector that helps to distinguish the two types and retaining only those that can be used for expression studies.
7. **Expression** – Once the transformed cells have been selected, they are analyzed for the expression (production of protein) of the desired gene or the fragment contained in the vector.

In the following discussion, these tools are described in brief. At the end of the section, all these tools are put together to demonstrate their roles in the construction of recombinant molecules. Of course, from time to time these primary tools have been supplemented by newer tools or methods which are also mentioned in brief.

Transformation Methods

The basic tools which are in use in biotechnology have their origin in the bacterial system. The natural transformation of DNA is very common among bacteria, which create chances for recombination to take place. The discovery of this feature led to the development of ways to induce transformation in laboratory conditions. Acquisition of foreign DNA by bacteria can be summarized in the form of three mechanisms, namely: **transformation, transduction and conjugation** (Fig. 1).

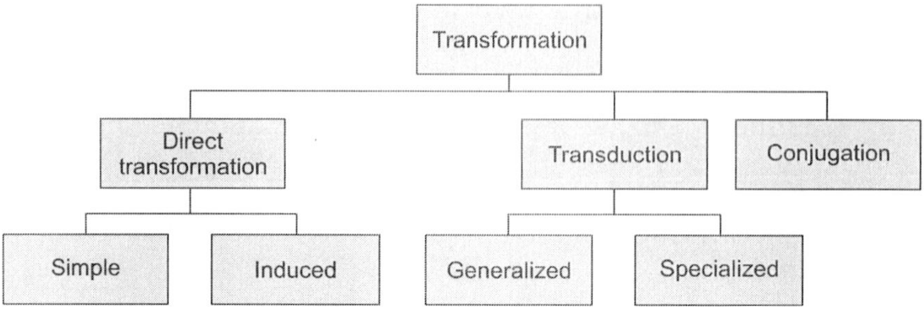

Fig. 1 Modes of transformation: direct transformation, transduction and conjugation.

(a) Transformation

(i) Simple Transformation

The process of DNA transformation was discovered in the late 1920's by Fred Griffith, an English medical officer, while experimenting with *Streptococcus pneumoniae*. Subsequently, this process of direct transformation of DNA was demonstrated in several bacteria. Direct transformation of DNA is the process in which exogenous DNA (from surrounding environment) is taken up by a (recipient) cell and is expressed. Although both eukaryotic and prokaryotic cells can take up DNA, but bacteria are highly transformable under normal growth conditions. However, not all bacterial cells can take up foreign DNA easily. Chemically treated cells that can take up DNA from the surrounding medium are called **competent cells**. Examples include species of *Bacillus, Haemophilus, Neisseria* and *Streptococcus* as their cells do not require special treatment to take up DNA and are naturally competent. In other bacteria, transformation is induced by specific chemical treatments.

There are many reports of natural transformation serving as a means of interaction among different groups of organisms. These even include the association of Gram-positive and Gram-negative bacteria, among yeast and bacteria, *Agrobacterium tumifaciens* and plants in which DNA from a plasmid in *Agrobacterium* gets integrated into plant chromosomes. Transformation thus can result in generating newer combinations or alteration of genetic make-up of bacteria.

(ii) Induced Transformation

The understanding of mechanisms of transformation led to the development of transformation methods. Initially, a method was developed to transform *Escherichia coli*, but transformation methods have been developed for variety of organisms and cells. One example of simple and traditional method of transformation is $CaCl_2$ heat shock treatment method (Fig. 2). In this method, overnight grown *E. coli* culture is pelleted

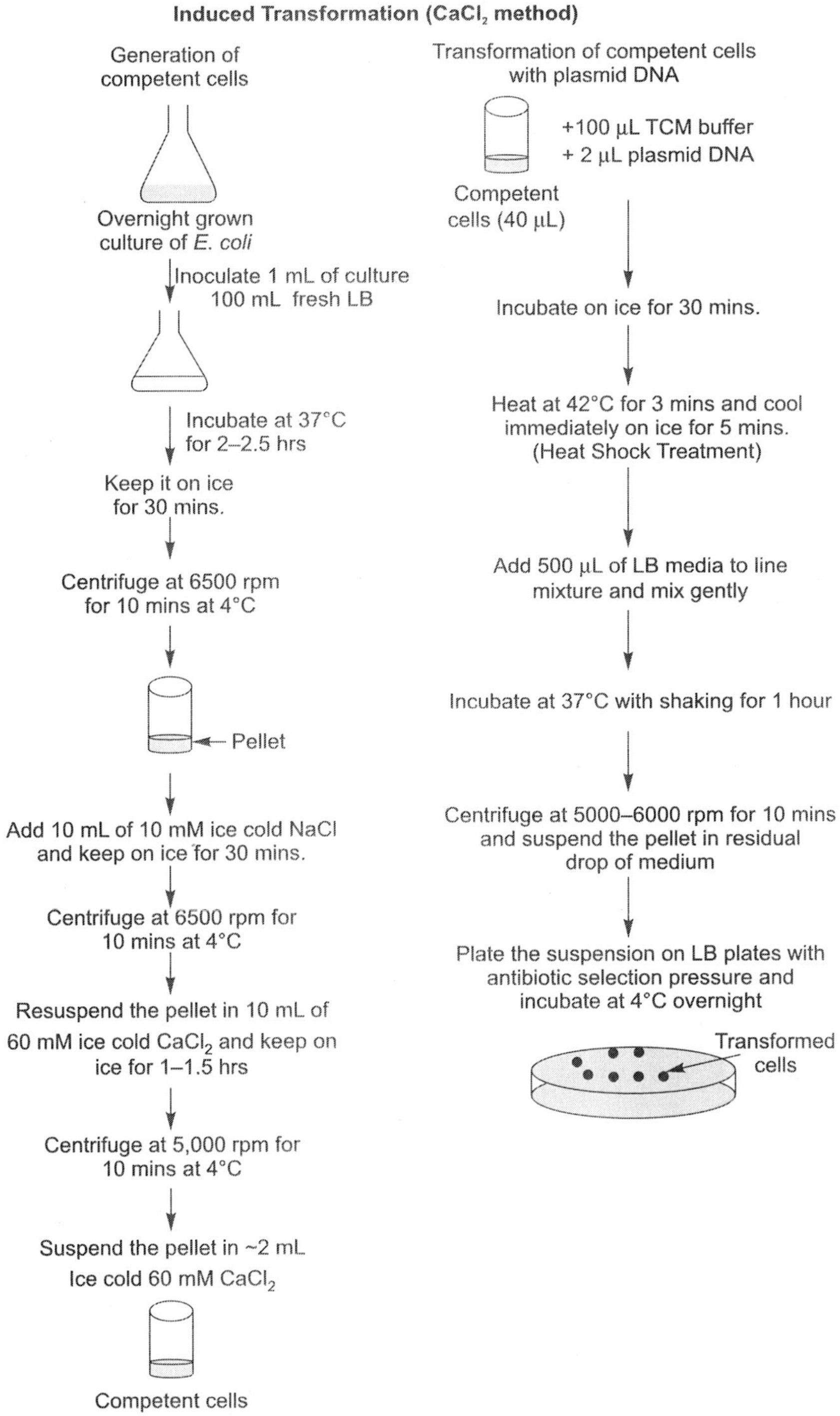

Fig. 2 Generation of competent cells in *E. coli* and transformation of plasmid DNA.

and treated with NaCl followed by suspension of the pellet in $CaCl_2$, and the suspension is kept on ice for 60–90 min. The pellet or the cells are then suspended in $CaCl_2$. The cells prepared in this way are known as the competent cells. Small aliquots of competent cells are taken to which plasmid DNA is added along with TCM (Tris-HCl, $CaCl_2$ and $MgCl_2$) buffer. DNA enters into the cells and then the transformants are selected in a medium containing antibiotic selection pressure, depending on the type of plasmid that has been used.

Another method called electroporation (Fig. 3) involves the use of an electric shock that temporarily breaches the cell membranes so that the DNA is taken up by cells. Both the above-mentioned techniques are discussed in Chapter 14 of this manual.

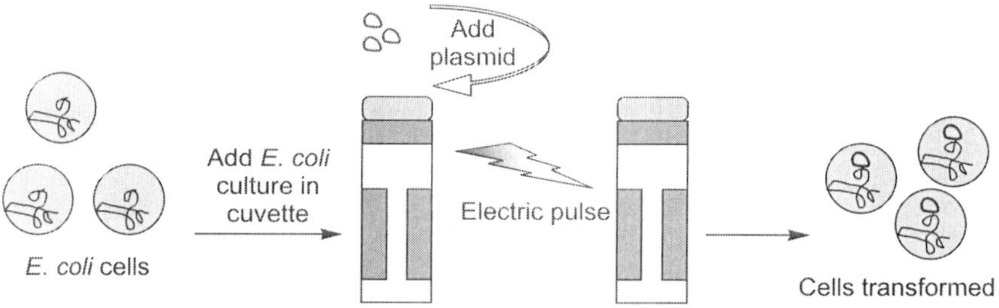

Fig. 3 Transformation in *E. coli* by electroporation.

(b) Transduction

Transduction involves the exchange of DNA among bacteria using bacterial viruses (bacteriophages) as intermediates. Transduction falls under two categories: **Generalized transduction** and **specialized transduction**. Both of them differ in their mechanism in the DNA that gets transferred. In order to explain transduction, it is necessary to understand the life cycle of a bacteriophage first.

When a bacteriophage infects a bacterial cell, it injects its DNA into the bacterial cell. The viral DNA is replicated numerous times, and viral genes are expressed by using the host cell ribosomes and other cytoplasmic components, producing the proteins that make up the viral capsid (or protein coat) and nucleases that digest the host genome into fragments. Simultaneously, the newly replicated viral DNA molecules are packaged into viral capsids, followed by lysis of bacterial cell (burst, and therefore killed). In this way, a progeny of hundreds of viral particles are released (Fig. 4).

(i) Generalized Transduction

Generalized transduction was first discovered by N.D. Zinder and J. Lederberg (1952). Although the infection and replication of a bacteriophage within its host is very precise, sometimes a mistake is made, and a fragment of the host (bacterial) DNA also gets packaged into a viral capsid. The phage so formed would be able to infect another cell, but it would not be able to replicate as it does not have any viral genes. The infected cell would have an extra piece of bacterial DNA, which could undergo recombination with the host chromosome, leading to a gene conversion event. This type of transduction is called generalized transduction because during this process, any bacterial DNA fragment can get packaged into the viral capsid (Fig. 5). Many genes are expected to have been acquired by bacteria through this process.

(ii) Specialized Transduction

Specialized transduction was discovered by Morse *et al.* in 1956. This type of transduction involves the transfer of only those chromosomal genes that are located adjacent to the attachment site of a prophage (bacteriophage genome with inserted and integrated into circular bacterial DNA). Only certain types of bacteriophage, such as phage **lambda** can act as a vector for specialized transduction to occur. Lambda has the ability to switch from **lytic cycle** to **lysogenic cycle** in a bacterial cell. In a lysogenic infection by lambda, integration of DNA takes place into a very specific spot in the host chromosome. The integrated viral DNA can remain integrated for long periods of time, without disturbing the cell (the details of regulation of phage infection and specialized transduction are not described here). Under the favorable conditions, the viral DNA will excise itself from the bacterial chromosome, and enter the **lytic phase**, in which the virus replicates just as described above and new bacteriophage particles are released when the host cell is lysed, to infect other cells. Sometimes the excision of lambda is disordered, and some bacteria DNA is excised along with it. When the resulting virus infects another cell, it passes on that bacterial DNA into the cell, along with its own DNA. If the infected cell survives (sometimes bacterial defenses against viral infection fails), it will contain a new piece of bacterial DNA, which can undergo recombination and possibly cause gene conversion. As long as the viral DNA integrates into a specific location, when it excises, the bacterial DNA removed with it will be the same in all cases. Therefore, the DNA transferred to the second cell will be the same segment of the bacterial chromosome. This is why this process is called **'specialized'** transduction (Fig. 6).

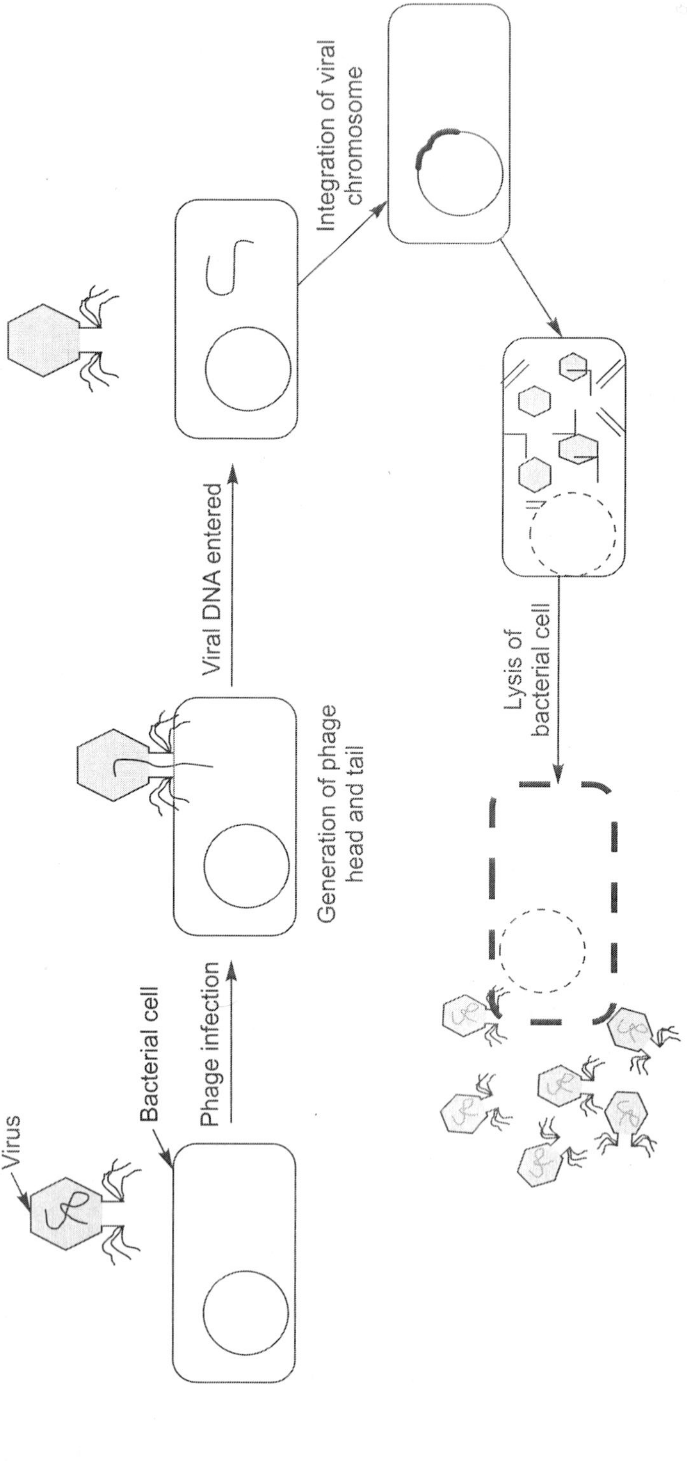

Fig. 4 Bacterial cells are lysed and viral particles are released in the surroundings.

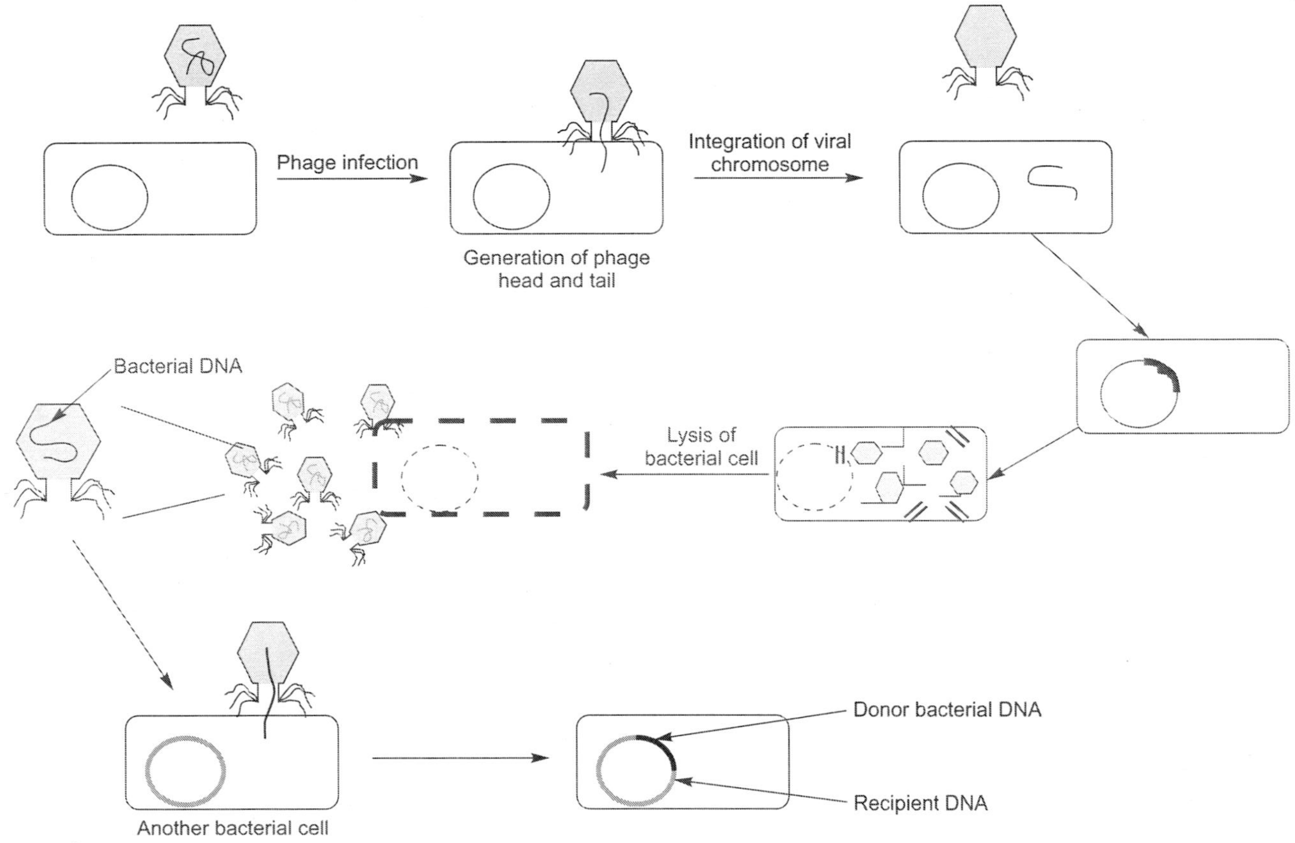

Fig. 5 Generalized transduction procedure showing transfer of any of the bacterial genes, facilitated by bacteriophages.

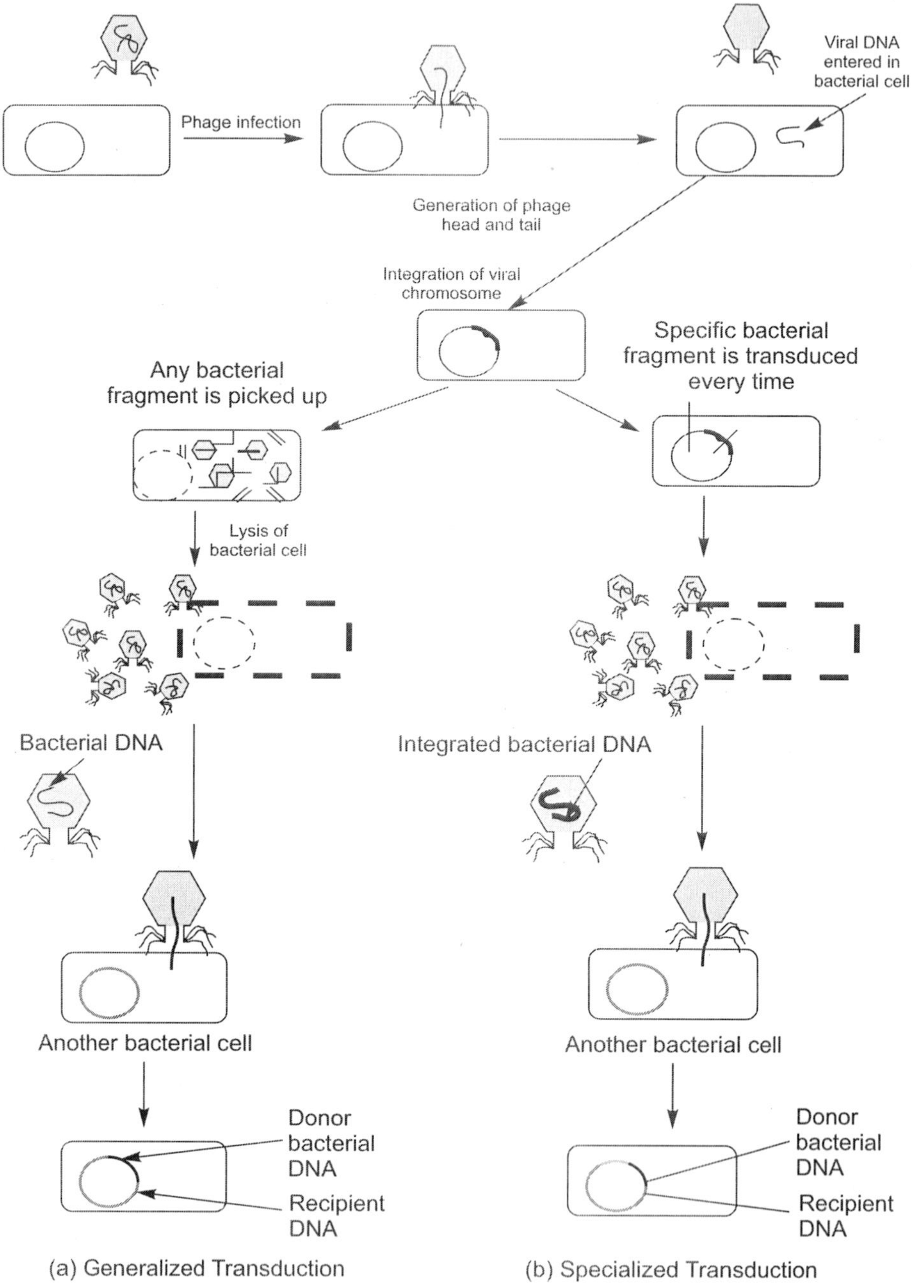

Fig. 6 Comparison of generalized and specialized transduction in bacteria.

(c) Conjugation

Conjugation was discovered by Joshua Lederberg and Edward Tatum in *E. coli* in 1946. It is the transfer of genetic material from one cell to another involving cell-to-cell contact (physical contact between bacteria). Conjugation is in fact a **mating** process among bacteria. The contact is made between the cells via a protein tube called an **F** or **sex pilus**, which acts as a conduit for the transfer of the genetic material.

Two strains of bacteria are involved in basic conjugation: strain with fertility factor (**F^+**) and strain without fertility factor (**F^-**). The F factor is an episome (episome is a circular piece of DNA that can replicate independently of the bacterial chromosome or integrate and replicate as part of the chromosome) that contains 19 genes and confers the ability to conjugate upon its host cell. The transfer of genetic material, i.e., the F factor in conjugation is from an F^+ cell to an F^- cell. The process of conjugation is depicted in Fig. 7.

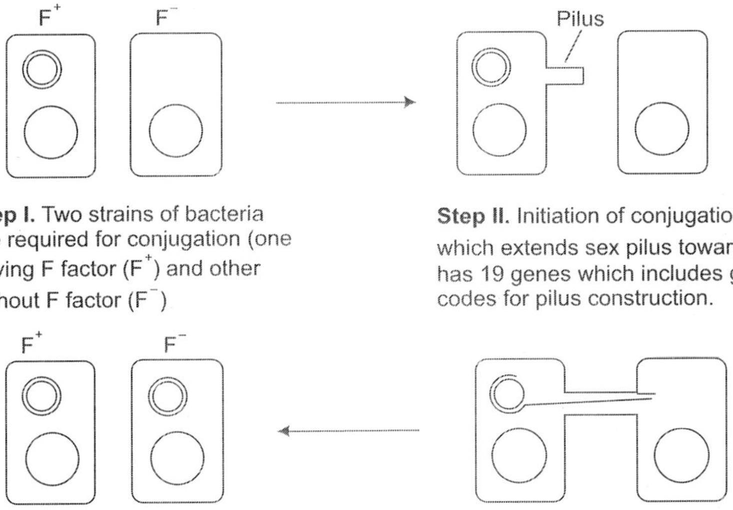

Step I. Two strains of bacteria are required for conjugation (one having F factor (F^+) and other without F factor (F^-)

Step II. Initiation of conjugation: F^+ cell which extends sex pilus towards the F^- cell has 19 genes which includes genes that codes for pilus construction.

Step III. This process produces a new F factor in the recipient cell. As the process gets over, the connection between the two cells is severed. Since both cells now contain an F factor, both cells are F^+.

Step IV. The F pilus temporarily connects the two cells. One strand of the F factor is nicked, and begins unwinding from the other strand. The nicked strand is then transferred through the F pilus to the F^- cell and strand replicating. The circular strand remaining behind in the F^+ cell also replicates.

Fig. 7 Process of conjugation between two bacterial cells belonging to same or different species.

Thus, it can be concluded that the evolution of bacteria not only depends on mutation and multiplication, but also on the transfer of DNA from one bacterial strain to another. This exchange is not limited from bacteria to bacteria, but from bacteria to other higher organisms also. Such transfer of gene is called **horizontal gene transfer (HGT).**

Existence of natural acquisition of DNA depends on chance factor. However, horizontal gene transfer does not happen so easily. Several factors act as hindrance to the process. The environment itself can lead to the degradation of naked DNA. Even if naked DNA succeeds in reaching the target cell, it can be mutated and degraded, or restriction enzymes can act on it as a defensive mechanism. By chance, if this DNA manages to escape such defensive mechanisms, then it can become a part of host DNA.

As a result of transformation, several genes and pathways have been evolved and subsequently spread among different organisms. For example, **Vancomycin** was an antibiotic of only hope against *Staphylococcus aureus* infections. Vancomycin resistance is associated with the **VanA gene** carried on a plasmid in other species of bacteria (such as *Enterococcus faecium* strain BM4147). The bad news is that *Staphylococcus* has been taking up DNA from other species that have overcome the commonly used antibiotics and learning how to survive by becoming resistant to most of the antibiotics used in hospitals. Recently, it was reported that there exists a StaphA strain **Mu50** that is showing signs of vancomycin resistance. Transformation of a plasmid from *Enterococcus* to *Staphylococcus* is the probable cause. Some other examples include resistance to several antibiotics, pathways of antibiotics biosynthesis, catabolic genes; opine production from bacteria to bacteria and from bacteria to plants (Table 2).

Table 2 Examples of Traits Acquired through Horizontal Gene Transfer among Bacterial Species

Trait	*Location in bacterial genome*			*Horizontal Gene Transfer*
	Transposon	*Plasmid*	*Chromosome*	
Resistance to antibiotics	+	+	+	B-B*
Production of antibiotics	+	+	+	B-B
Insecticidal toxins	–	+	–	B-B, B-P**
Metabolic pathway	–	+	+	B-B, B-Y***
Degradation of halogenated organic compounds	+	+	+	B-B
Opine production	–	+	–	B-B, B-P

*B-B----natural process of transformation, **B-P----engineered gene between bacteria and plants, ***B-Y----engineered genes between bacteria and yeast

CLONING VECTORS

The second factor or tool which is simultaneously responsible for the rapid growth of biotechnology is the availability of cloning vectors. A cloning vector designed for a microorganism is an agent that facilitates the overall process of placing foreign genetic material from any source into the recipient. In fact development of cloning vector and transformation methods are interdependent. You cannot develop a transformation method without a cloning vector and cloning vector cannot be developed without a transformation method. Cloning vectors are generally derived from either plasmids or viruses (Fig. 8).

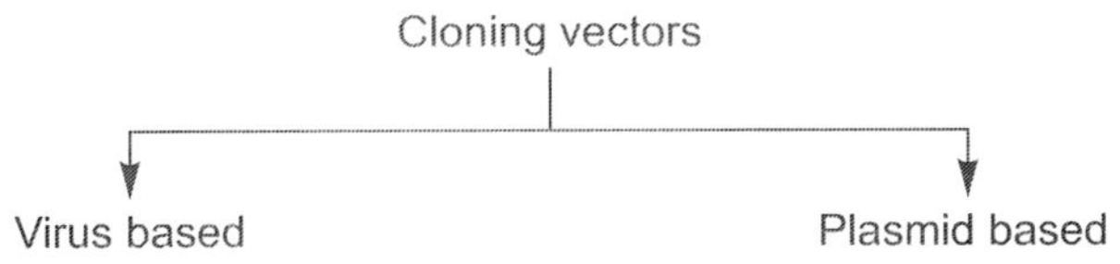

Fig. 8 Cloning vectors are categorized as plasmid based and virus based cloning vectors.

Plasmids

Most of the cloning vectors initially were developed from plasmids. The term "plasmid" was coined by Joshua Lederberg in 1952 and since 1969; this term is in frequent use. Generally, plasmids are very common among bacteria, although they have also been reported from fungi. The size of natural plasmids varies from 1kb to more than 200 kb. Although, plasmids are extrachromosomal, double stranded covalently closed circular DNA molecules, but linear plasmids have also been reported. They replicate along with the replication of bacterial chromosome and are inherited independently. But they are dependent on the host for most of the functions like replication and transcription.

Plasmids are known to contain genes that are associated with resistance to antibiotics, antibiotic production, degradation of complex and recalcitrant compounds, restriction enzymes and several other products. It is interesting to note that entire plasmid DNA is not needed for plasmid maintenance in the host. A specific portion of plasmid DNA called origin of replication (*ori*) or plasmid replicon (***rep***) is all that is essential for its maintenance in a specific host. A plasmid replicon is generally defined as the smallest piece of plasmid DNA that is able to replicate autonomously and maintain the copy number, e.g., pA387 (Fig. 9), pUC 18/19.

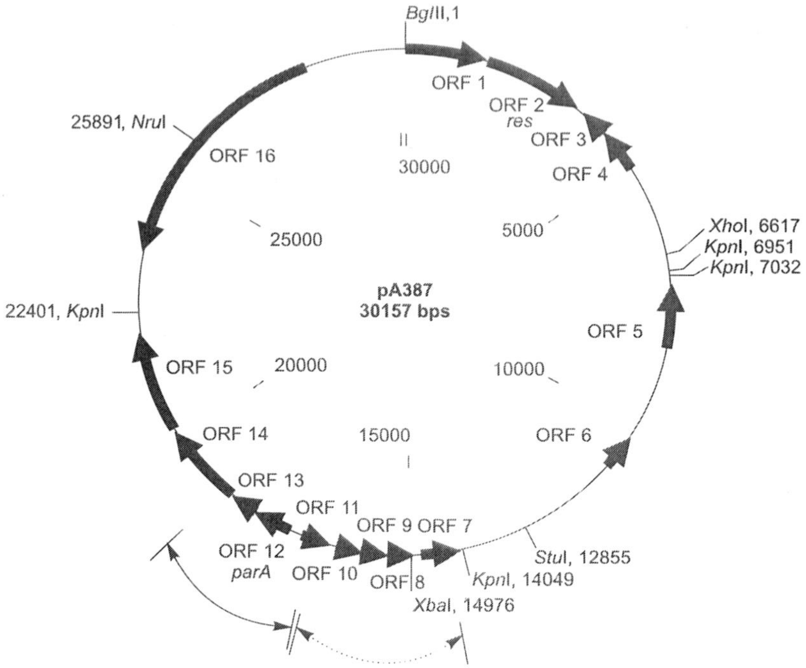

Fig. 9 Physical map of plasmid pA387 from Amycolatopsis benzoatilytica DSM 43387. The map positions of selected restriction sites on pA387 are indicated (shown in base pairs). Solid outer arrow indicates the stability region of pA387. Dotted outer arrow indicates the replication region of pA387.

Plasmids are host specific, i.e., a plasmid capable of replicating in *E.coli* need not necessarily replicate or stay itself in *Arthrobacter*.

Thus, in case we are interested in genetic manipulation of *Arthrobacter* that would mean that we need to develop cloning vector independently for *Arthrobacter*.

Since a plasmid is (by definition) an extrachromosomal element, therefore it cannot make use of any origin of DNA replication present in a chromosome. That is, origin of DNA synthesis (*ori*) is required for DNA replication within (i.e., copying of) a plasmid. Irregularity in plasmid replication would lead to dilution of that plasmid in a population of dividing cells because it couldn't be passed on to daughter cells.

Within a cell, a plasmid can exist in many forms, i.e., linear, supercoiled, circular and polymorphic forms and these forms can be visualized on gel as shown in Fig. 10. Due to conformational constraints, supercoiled DNA is more compact in shape than relaxed DNA of same molecular weight. Therefore, when run on a gel, supercoiled DNA moves faster than the relaxed DNA and later moves faster than open circular DNA.

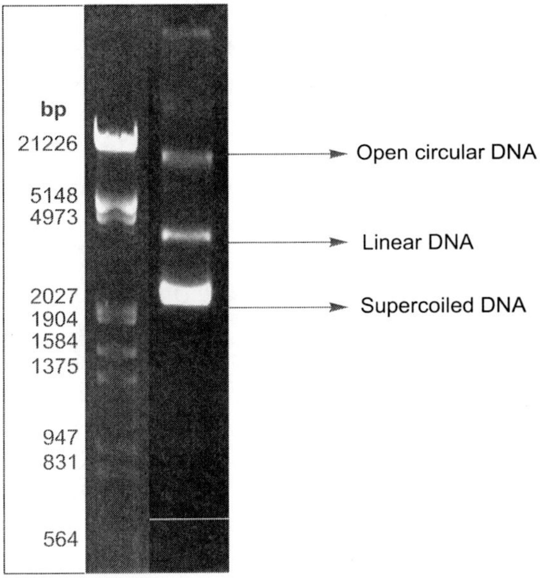

Fig. 10 0.8% gel electrophoresis showing mobility of supercoiled, linear and open circular DNA. Lane 1: Lambda *Hind*III/*Eco*RI double digested DNA (Fermentas); Lane 2: Plasmid DNA pUC18.

Plasmid based cloning vectors

Unnecessary segments from natural plasmids are removed and essential sequences are added to generate **plasmid cloning vectors**. To clone DNA fragments, the same restriction enzyme must be used to cut both the vector and the DNA sample, so that they have similar ends. Therefore, vector usually contains a sequence (polylinker) which can recognize several restriction sites and the vector can be used for cloning a variety of DNA fragments. A plasmid vector should also contain marker genes (generally antibiotic resistant genes) for selecting the transformants. The vector after entering a host cell may proliferate with it. However, most *E. coli* cells that proliferate in the medium would not contain any plasmid because the transformation efficiency of the plasmid in *E. coli* is very low. Thus, antibiotic resistant genes are used as **selective markers** since antibiotic is added to the medium which kills the cells that do not contain any vector.

A typical cloning vector should have the following properties:

- Minimum size
- A selectable marker gene (e.g., resistance to ampicillin)
- An origin of replication (*ori*)
- Multiple cloning site (a place to insert foreign DNA or polylinkers)

Looking into the requirements of an ideal cloning vector, we need to add one or two marker genes. Why do we need marker genes? For the reason that when we do the transformation, not all but only a few cells (say one out of 10) will take up the plasmid DNA. Thus, we need to eliminate transformants from non-transformants and this can be done by using a **selection pressure** or by using any phenotypic character on the plasmid.

A selectable marker is a non-essential element of a plasmid, but enables us to maintain those cells that contain the plasmid uniformly. Sometimes, carrying a plasmid puts a cell at a ***selective disadvantage*** compared to its plasmid-free neighbors, so the cells with plasmids grow more slowly. Cells that happen to "kick out" their plasmid during division may be **"rewarded"** by having a higher rate of growth, and as a result these plasmid-free (sometimes referred to as "cured") cells may take over a population. But if a plasmid contains a gene that the cell needs to survive (for example, a gene encoding an enzyme that destroys an antibiotic), then cells that happen to kick out a plasmid are **"punished"** (by subsequent death) rather than "rewarded" (as in the previous scenario). Thus, selective pressure helps to maintain a plasmid in a population.

Initially, during early 1970s, selectable markers like Kan^r (kanamycin resistant gene), Amp^r (ampicillin-resistant gene), tet^r (tetracycline-resistant gene) were introduced into plasmids having pMB1 or colE1 replicon and first plasmid cloning vectors appeared were pSC101, ColE1 and pCR1. However, none of them carried more than two restriction sites suitable for cloning. The first cloning vector that contained pMB1 replicon and two antibiotic resistance genes, tet^r and amp^r, was pBR313. However, this plasmid was larger in size and contained some extra DNA which made it bulkier. Thus, a derivative of pBR313 with a size of 4.36 kb having additional features for cloning was developed. The plasmid was named pBR322 (Fig. 11).

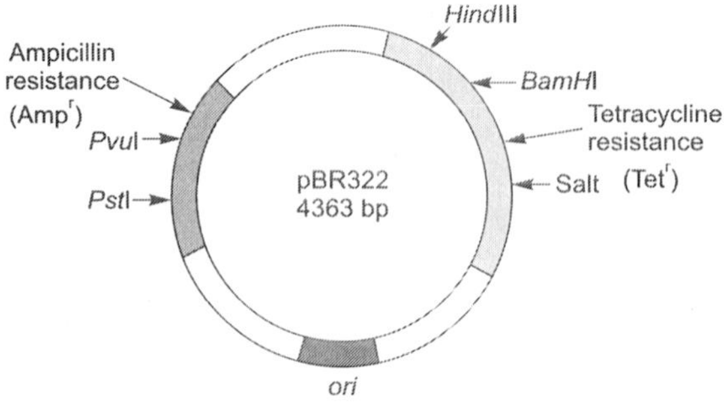

Fig. 11 Restriction map of pBR322 showing origin of replication (*ori*), selectable markers (Amp^r and Tet^r) and few important restriction enzymes. Note that *Pst*I and *Pvu*I (within Amp^r) and *Hind*III, *BamH*I and *Sal*I (within Tet^r) lie within marker genes and can be used for cloning.

Until 1978, pBR322 was the only cloning vector for use. However, subsequent five years saw a rapid progress in the improvement and development of better cloning vector. Major change was reduction in size that led to accommodation of larger fragments of DNA and also, improvement in transformation efficiency. In addition, smaller size also increased the copy number. Unfortunately, the first generation high copy number plasmids such as pXF3 and pAT53 which were developed suffered from major drawback as there were not many restriction sites located within them. Thus, these plasmids were soon replaced by a series of cloning vectors commonly called **pUC** vectors (Fig. 12).

In the pUC vectors, synthetic cloning sites (also known as polylinkers or multiple cloning sites) were introduced. The polylinkers contain sequences that contain series of restriction enzymes. These restriction sites are generally unique as they are not located anywhere else within plasmid.

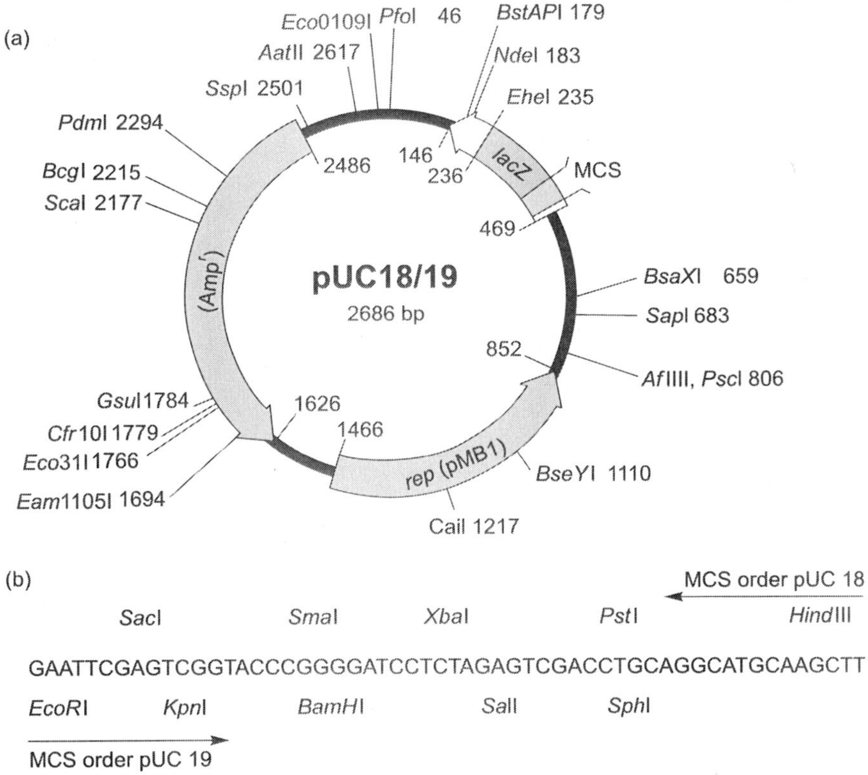

Fig. 12 (a) Restriction map of plasmid based cloning vector pUC19 showing multiple cloning site inserted (MCS) within LacZ gene, a selectable marker (amp^r gene) and region of origin. (b) Sequence of MCS showing various Restriction Enzymes present. These enzymes can be selectively used to clone a foreign DNA fragment. The MCS in pUC19 is reverse orientated to those of the pUC18.

Although, insertional inactivation is not possible with pUC vectors as they carry only one antibiotic resistance gene that confers resistance to ampicillin (amp^r). However, amp^r is used only for the selection of the transformants as pUC vectors contain segments of *E. coli* DNA that contains the regulatory sequence of lacZ gene and the coding information of 146 amino acids of β-galactosidase.

Downstream to the initiation codon ATG, a multiple cloning site is embedded. The 146 amino acid β-galactosidase expressed by pUC in *E. coli* does not have any β-galactosidase activity. However, *E. coli* strains that have gene on the chromosome, produce the second unit of the enzyme and together these two chains form the active β-galactosidase. Conversely, cloning into the polylinker inactivates the α-fragment of pUC and thus active β-galactosidase is not produced in the recombinants. This activity can be easily detected by growing the transformant cells in a medium containing a chromogenic substrate 5-bromo-4-chloro-3-indole-β-D-galactoside (X-gal) which when acted upon by β-galactosidase produces an intermediate that imparts blue color to the colonies. Thus, those transformants in which α-fragment of the lacZ gene is inactivated appear as white colonies in contrast to the normal transformants that appear blue in color (Fig. 13).

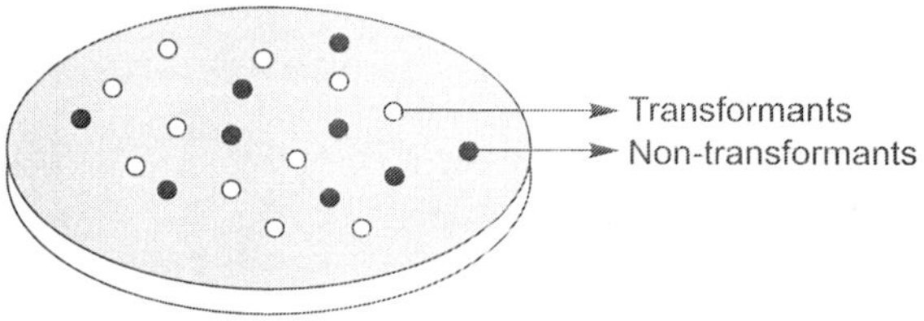

Fig. 13: Cloning in MCS site of pUC vector leads to inactivation of α chain of β-Galactosidase. Therefore, white colonies are formed as β-Galactosidase cannot act upon X-gal and therefore, no blue color is formed in transformed colonies.

When the cells are spread on a bacteriological plate lacking antibiotic, a solid lawn of bacteria covering the plate is seen the next day as very few cells take up DNA and most cells are usually killed by the antibiotic in a typical transformation experiment. Sometimes, a few colonies survive on the plate and these are generally mutants. When an agar plate containing an antibiotic and transformed colony is kept for a longer duration, the real colonies are destroyed.

On the other hand, some cells do survive because other cells are doing the work of destroying the antibiotic in their immediate vicinity on the plate. These are called "satellite colonies" (Fig. 14) because they are tiny little colony specks that surround the big antibiotic-resistant colony. Such colonies develop with antibiotics, such as ampicillin, which are destroyed by enzyme β-lactamase secreted outside of the cell. These satellite colonies do not grow when plated on a medium containing antibiotics, so they are not a problem. Certain conditions are also favorable for the development and growth of satellite colonies like when ampicillin plate becomes old (meaning that the antibiotic is partially degraded), or the transformed cells are plated at very high density (meaning that the plate is covered with resistant cells), or the copy number of the plasmid in the cells is so high that β-lactamase is secreted at high levels, or the colonies grow on the plate for several days (allowing more time for degradation).

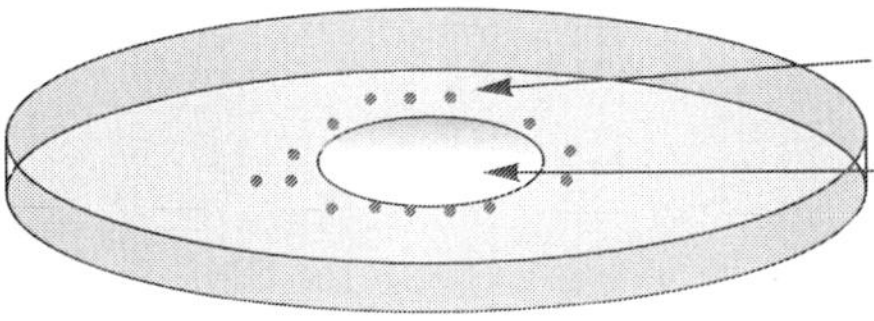

Fig. 14 The big colony here is an antibiotic-resistant colony that is surrounded by many satellite colonies. While big antibiotic colony is doing its job of destroying antibiotics, such satellite colonies get a shelter around it and grow in antibiotic-free environment.

Bacteriophage vectors

The mode of action of bacteriophage is already explained above (transduction). The most widely used bacteriophage vector is the one which infects *E. coli* as it possesses many advantages because of its high efficiency with which it can infect *E. coli* and its extraordinary ability to take up fragment of nearly 20 kb. The essential genes of the phage genome are clustered together. The strain phage Lambda (λ) has been engineered by incorporating two *Eco*RI sites on either side of the non-essential genes, so that the non-essential genes can be removed by using *Eco*RI. Foreign DNA cut with *Eco*RI will generate sticky end complementary to those of *Eco*RI digested λ. The two types of DNA are cut and ligated (Fig. 15). The **cos** sites present on the λ phage are required for packaging. The recombinant molecules are then packaged and transfected into *E. coli*. The infected places can be detected by the formation of plaques.

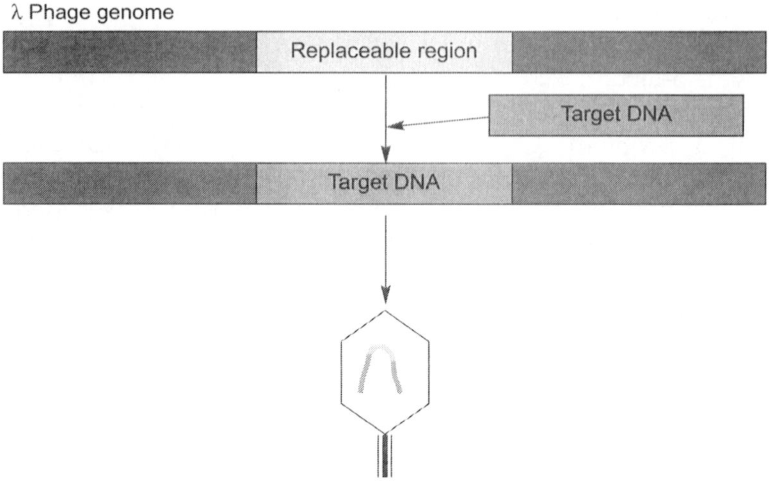

Fig. 15 Cloning of DNA using λ vector. Non-essential DNA can be replaced by foreign DNA. This foreign DNA along with the viral DNA is packaged by *in vitro* assembly to get the recombinant DNA molecule.

Cosmid vectors

These carry the **cos** sites of phage λ; they can be packaged into viral coats and transferred into bacteria by viral infection. A cosmid can carry more than twice as much DNA as can a phage vector and all viral genes are deleted (except cos sites). Cosmid vectors have the following important features:

1. A plasmid origin of replication (*ori*)
2. One or more unique restriction sites
3. One or more selectable marker genes
4. Cos sites to allow the packaging of DNA into phage heads

Cosmids behave like a plasmid, but recombinant molecules are first transferred through transfection into *E. coli* where the cosmids replicate like plasmids (Fig. 16).

Specialist vectors

The vectors described above are good for generating libraries, but there are specialist vectors also which are used for expressing the foreign insert DNA in bacterial cells. These include special vectors like:

(a) **Expression vectors –** These vectors enable cloning of genes under the control of a promoter that can function in *E. coli* cells (protein expression). e.g., pGEX-3x (bacterial cells).

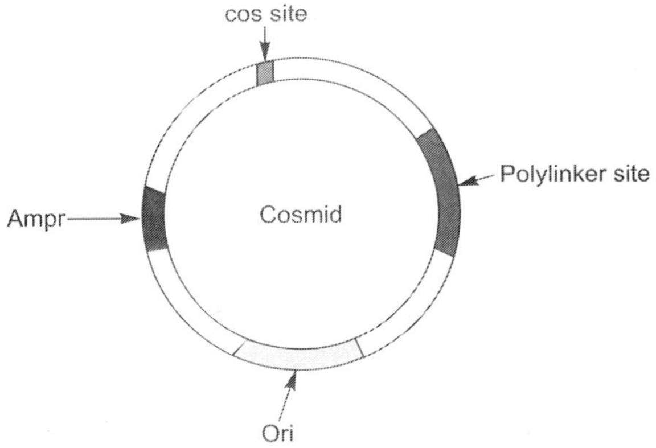

Fig. 16 Cosmid vector with cos sites of phage λ, plasmid origin of replication.

(b) **Shuttle vectors** - These are plasmid vectors that have been designed to contain two origins of replication. This allows them to replicate in two different hosts like E. coli and Streptococcus.

Innumerable cloning vectors have now been constructed (Table 3).

(a) **Plasmid**—they are self-replicating extrachromosomal bacterial elements.

(b) **Bacteriophages or bacterial viruses** like M13, lambda and P1, etc.

(c) **Cosmids and fosmids**—They are combination of plasmid and bacterial virus DNA sequences. Fosmids are similar to cosmids, but they are based on F-Factor of bacteria.

(d) **Bacterial artificial chromosome (BAC)** - It is derived from a plasmid that exists as a single copy in *E. coli* and is meant to carry large DNA insert.

(e) **Yeast and human DNA sequences** including a site for the initiation of DNA synthesis (origin of replication, *ori*) as well as centromere and telomere sequencing for constructing artificial chromosomes.

Attributes of a cloning vector:

1. Their replication and copy number is maintained in the host cell.
2. Presence of unique restriction sites providing cloning sites for inserting source DNA.
3. Selectable marker genes to select the recombinants from non-recombinants.

Table 3 List of the Vectors Used in their Respective Host System with Their Insert Size Capacity

Vector System	*Host system*	*Insert capacity (kb)*
Plasmid	*E. coli*	01–10
Bacteriophage lambda	Lambda/*E. coli*	10–20
Cosmid	*E. coli*	35–45
Bacteriophage P1	*E. coli*	80–100
Bacterial artificial chromosome (BAC)	*E. coli*	50–300
P1 derived artificial chromosome	*E. coli*	100—300
Yeast artificial chromosome (YAC)	Yeast	200—2000
Human artificial chromosome (HAC)	Cultured human cells	>2000

ENZYMES USED IN MANIPULATION OF DNA

The skill to manipulate DNA molecules in laboratory conditions forms the basis of recombinant technology. Manipulation of DNA molecules depends on the accessibility of enzymes performing specific activities that can be controlled. These enzymes primarily include the following:

- Restriction endonucleases
- DNA polymerases
- Nucleases
- Ligases
- End-modification enzymes

1. Restriction Endonucleases

Of all the enzymes, restriction enzymes are of primary importance. They are also known as restriction endonucleases as they cut within the molecule. These **DNA-cutting enzymes** are found in bacteria as a mechanism for self-defense and protect them by digesting any incoming foreign DNA. There are four major types of restriction endonucleases designated as Type I, Type II, Type III and Type IV. Of these, Type I and Type III restriction enzymes do not always cut at specific sites and hence, lose their importance in molecular biology. The type II restriction enzymes recognize a specific sequence in the DNA molecules and make two cuts one in each strand (Fig. 17 creating either blunt ends or cohesive ends. Considering the reproducibility in the patterns of fragments obtained, restriction endonucleases have been exploited for the generation of recombinant DNA molecule. Currently, more than 1000 restriction endonucleases are available.

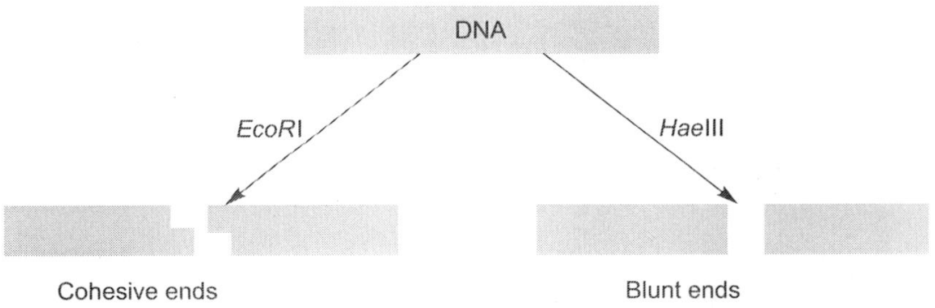

Fig. 17 Restriction enzymes cleave the DNA at specific sites and either produce cohesive or blunt ends. This pattern of cleaving is specific for respective enzymes.

Restriction enzymes are conventionally named by their host of origin. Say for instance, the restriction enzyme *Eco*RI is derived from **E**scherichia **co**li, *Hind*III from **H**aemophilus **in**fluenzae, etc. Restriction enzymes cleave DNA only at a specific sequence which they recognize is known as the recognition sequence. For example, an enzyme named ***Hae*III** produced by *Haemophilus aegyptius* recognizes the sequence 5′GGCC3′. Wherever this sequence is found in DNA molecule, the enzyme cuts between G and C. This recognition sequence is present at 11 locations in viral genome of fX174, therefore when treated with *Hae*III, 11 fragments of different lengths are produced.

Restriction-modification systems

Restriction modification systems comprises the combination of restriction endonuclease and methylases which bacteria employ for their defense purpose. Restriction enzymes and methylases work mutually within a bacterium. The restriction enzymes detect the foreign DNA and cleave it into fragments. Therefore, the foreign DNA is degraded. But if the foreign DNA can be degraded by methylases alone, then what are methylases for? The answer lies in the simple fact that restriction enzymes cut DNA at specific site irrespective of whether the DNA belongs to the host or virus. And as a result, the bacterial DNA needs to be modified in such a way that they escape from these restriction endonucleases. So, the bacterial DNA is methylated in order to protect its own genomic DNA from cleavage.

Restriction enzyme recognition sequences

As explained above, the restriction enzymes cleave at specific sites (called restriction sites) of double-stranded DNA. These sites may vary from 4 bps to even exceeding more than 8 bps. For example, enzymes *Eco*RI, *Sac*I and *Sst*I recognize 6 bp restriction sites of respective sequences, whereas Sau3AI recognizes 4 bp sequences. The length of restriction site is inversely proportional to the frequency with which these restriction sites occur within the genome. Enzymes with a 6 bp recognition site will cut, on average, every 4^6 or **4096 bp**; a 4 bp recognition site will occur roughly every 256 bp.

Different restriction enzymes can have the same recognition sequence known as **isoschizomers.** These isomers can vary in their physiological characteristics like optimum conditions, stabilities, etc. It may be noted that although the two restriction enzymes can share the same recognition site, but the pattern of chopping the DNA may vary.

Some restriction enzymes are very specific while cleaving the DNA. Variation in even a single base of restriction site modifies their activity. Such recognition sites are called **unambiguous sites**. For example, *BamH*I recognizes the sequence GGATCC only. But enzymes like *Hin*fI can recognize a 5 bp sequence starting from GA and ending at TC, no matter which nucleotide falls in between! Therefore, *Hinf*I has ambiguous restriction site. **The recognition site for one enzyme may contain the restriction site for another.**

Another point to ponder is that most of these restriction enzymes are palindromes, i.e., they read the same forward and backward. Enzyme *Eco*RI acts on the palindromic sequence being 5'GAATTC3' on the top strand and 5'CTTAAG3' on the bottom strand. This may be noted that most but not all restriction enzymes recognize such palindromic sequences.

Patterns of DNA cutting by restriction enzymes

This is a well-known fact that phosphate bonds form the backbone of DNA. Restriction enzymes hydrolyze this backbone, i.e., they cut between deoxyribose sugar and phosphate groups. Therefore, a phosphate group is set free on the 5′ ends and a hydroxyl group on the 3′ ends of both strands. Restriction enzymes usually cut in three patterns which produces either 5′ overhangs, 3′ overhangs or blunt ends. 5′ overhangs and 3′ overhangs so produced by these ezymes are also known as sticky ends or cohesive ends because they anneal very easily with their partners.

5′ Overhangs: An asymmetrical cleavage within the recognition site produces a short single-stranded segment extending from the 5′ ends. For example, *BamHI* enzyme cuts in the following manner.

5′-G A A T T C-3′
3′-C T T A A G-5′

↓ *EcoRI* cleavage

5′-G/A A T T C-3′
3′-C T T A A/G-5′

↓

5′-G 5′-A A T T C-3′
5′-C T T A A-5′ G-5′

3′ Overhangs: An asymmetrical cleavage within the recognition site can also result in overhangs from the 3′ ends of both the strands. For example, *Kpn*I produces 3′ overhangs as shown below.

5′-C T G C A G -3′
3′-G A C G T C -5′

↓ *Pst*I cleavage

5′-C T G C A/G -3′
3′-G/A C G T C -5′

↓

5′-C T G C A -3′ G -3′
3′-G 3′-A C G T C-5′

Blunt ends: When an enzyme cleaves the recognition sequence symmetrically, it forms blunt ends in spite of producing overhangs. For example, *Sma*I produces blunt ended fragments by cleaving its recognition sequence.

5′- C C G C G G -3′
3′- G G C G C C -5′

↓ *Sma*I digestion

5′- C C G/C G G -3′
3′- G G C/G C C -5′

↓

5′-C C G C G G-3′
3′-G G C G C C-5′

2. DNA Polymerases

Synthesis of DNA *in vitro* is essential for many techniques like PCR, DNA sequencing, DNA labeling and many other procedures vital to research in molecular biology. DNA polymerases are used to synthesize new DNA from existing template; hence it is also known as template-dependent DNA polymerase.

The mode of action of a template-dependent DNA polymerase: It synthesizes new strand of DNA complementary to the template being copied, therefore it also acts in a similar way like template-dependent RNA polymerase. The new polynucleotide is synthesized in 5′–3′ direction which resembles the *in vivo* mechanism of synthesis of DNA molecule (Fig. 18). But one important difference lies in the fact that the DNA polymerase requires an initiator to start the synthesis of the complementary DNA strand. Therefore, a primer (17–20 nucleotides long which is complementary to the template strand) is added in the reaction mix. The primer anneals with the template strand where it finds its corresponding sequences, and then DNA polymerase starts its job of synthesizing the rest of the strand. As the primer is specific for its binding region, therefore, the required region of DNA can be synthesized by designing the primers suitable for that region. The procedure of designing primers has been illustrated in detail in chapter 9 of this manual.

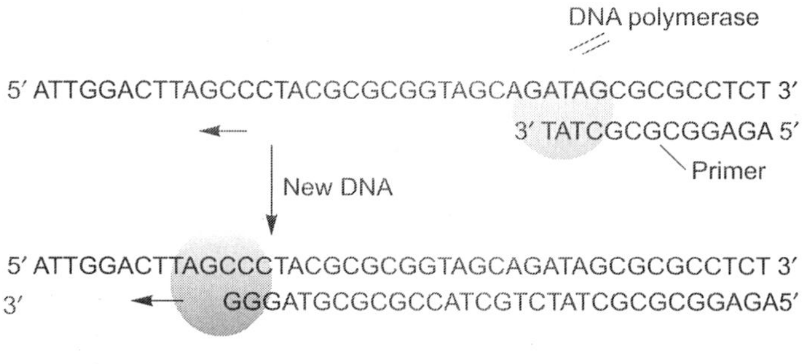

Fig. 18 Mode of action of DNA polymerase. A copy of template DNA is synthesized in the 5′-3′ direction which is complementary to it.

But the role of template-dependent DNA polymerases does not end here. In addition to the polymerization ability, DNA polymerases may show 5′→3′ or 3′→5′ exonuclease activity or both of the following:

3′→5′ exonuclease activity- This exonuclease activity removes the erroneous nucleotides from 3′→5′ end of the newly synthesized DNA strand. Therefore, this activity is also known as **proofreading** activity.

5′→3′ exonuclease activity- This activity is less common, but still exists in nature in some of the DNA polymerases so that at least a part of the already synthesized incorrect polynucleotide can be removed.

Types of polymerases

1. **Kornberg Polymerase:** Kornberg polymerase is a template-dependent DNA polymerase, which was named after its discoverer Arthur Kornberg. This is one of the versions the *Escherichia coli* DNA polymerase I enzyme. This enzyme has both 3′–5′ and 5′–3′ exonuclease activities and therefore, is primarily used in DNA labeling.

   The 3′–5′ exonuclease activity is always better for an enzyme for editing or proofreading. But a problem arises when the enzyme possesses 5′–3′ activity too. Such enzymes can *in vitro* remove nucleotides from 5′ end of a newly synthesized DNA strand. Such modifications may not degrade the complete strand, but can create problems in the long run. Especially, when we talk about PCR, such modifications for a few cycles can create a lot of difference. Also, in case of sequencing, 5′ end modification may not lead to correct DNA sequence. Hence, there was a need to search for such enzymes in molecular biology that do not have 5′–3′ exonuclease activity.
2. **Klenow Polymerase:** This is a modified version of Kornberg polymerase. For sequencing and other experiments in molecular biology, naturally existing *E. coli* polymerase I was modified by cutting it into two segments with protease. After protease activity, the polymerase retained 3′–5– exonuclease activity, but was devoid of 5′–3′ exonuclease activity. But this enzyme also could not be used for sequencing and was confined for the use of end labeling only.
3. **Taq-DNA polymerase:** This was isolated from a thermophilic bacterium named *Thermus aquaticus*, originally isolated by Chien *et al.*, in 1976. The polymerase is very stable at high temperature and therefore has application in polymerase chain reaction. The optimum temperature for its activity is between 75 and 80°C and can replicate a 1000 base pair strand of DNA in less than 10 seconds at 72°C. But it lacks of 3′ to 5′ exonuclease proofreading activity resulting in relatively low replication fidelity.
4. **Sequenase:** This enzyme was discovered in 1980s and was proved to be an escort in the field of molecular biology. It is a genetically engineered form of T7 DNA polymerase which lacks 3′→5′ exonuclease activity. It has the ability to incorporate nucleotide analogs (dITP, thio-dNTPs, dideoxy-NTPs, etc.), and can carry out strand displacement synthesis. It is an excellent enzyme for dideoxy-sequencing, and is also useful in other applications, especially where the absence of associated exonuclease activity is desirable.

5. **Reverse Transcriptase:** It is a RNA dependent DNA polymerase enzyme which synthesizes DNA from RNA. These enzymes are important in RNA viruses' replication cycle while infecting host cells. But, in molecular biology, these enzymes are used to make copies of DNA from mRNA, i.e., complementary DNA (cDNA) is produced from mRNA. The cDNA has advantage over genomic DNA in lacking nucleotides that code for introns.

3. Nucleases

Another class of enzymes used in recombinant DNA technology are nucleases which cleave off nucleotides from nucleic acids (DNA/RNA). Majorly, nucleases are of two types, namely, **exonucleases** (removes nucleotides from the ends of DNA/RNA molecules), and **endonucleases** (cleaves internal phosphodiester bonds in DNA). Some of them are DNA specific while others possess specificity for RNA. Few nucleases act only on double-stranded DNA, whereas others cleave single-stranded DNA molecule.

4. DNA Ligases

There are various ways to create a recombinant DNA molecule. Recombination includes generating new combination of DNA either by cutting or joining. Ligation means joining of linear fragments together with covalent bonds. Covalent bond formed in ligation reaction is a phosphodiester bond between 3'- OH of one nucleotide and 5' phosphate of the next. The most widely used DNA ligase is derived from the T4 bacteriophage which requires ATP (adenosine triphosphate) as a cofactor whereas DNA ligase obtained from *E. coli* utilizes NAD as a cofactor.

An enzyme called T4 DNA ligase is used to anneal DNA fragments possessing overhangs, **sticky ends** as shown below (Fig. 19). This is equivalent to repairing "nicks" in duplex DNA. Using the higher concentration of this enzyme blunt ended DNA fragments can also be ligated. An elaborate account of the procedure to perform this reaction has been given in Chapter 13.

Three components in addition to water are required for ligation reaction:

- Blunt ended or sticky ended fragments of DNA.

An ATP containing buffer (stock prepared as 10X), working dilution of buffer used is 1X yielding about 0.25 mM to 1 mM of ATP in reaction. **T4 DNA ligase** for ligating a fragment into a plasmid vector. Ligation of sticky ends requires about 0.01 units of enzyme, whereas blunt ends utilize about 1 unit of ligase enzyme. Temperature and incubation time together dictate the activity of ligase, for example, ligations carried out

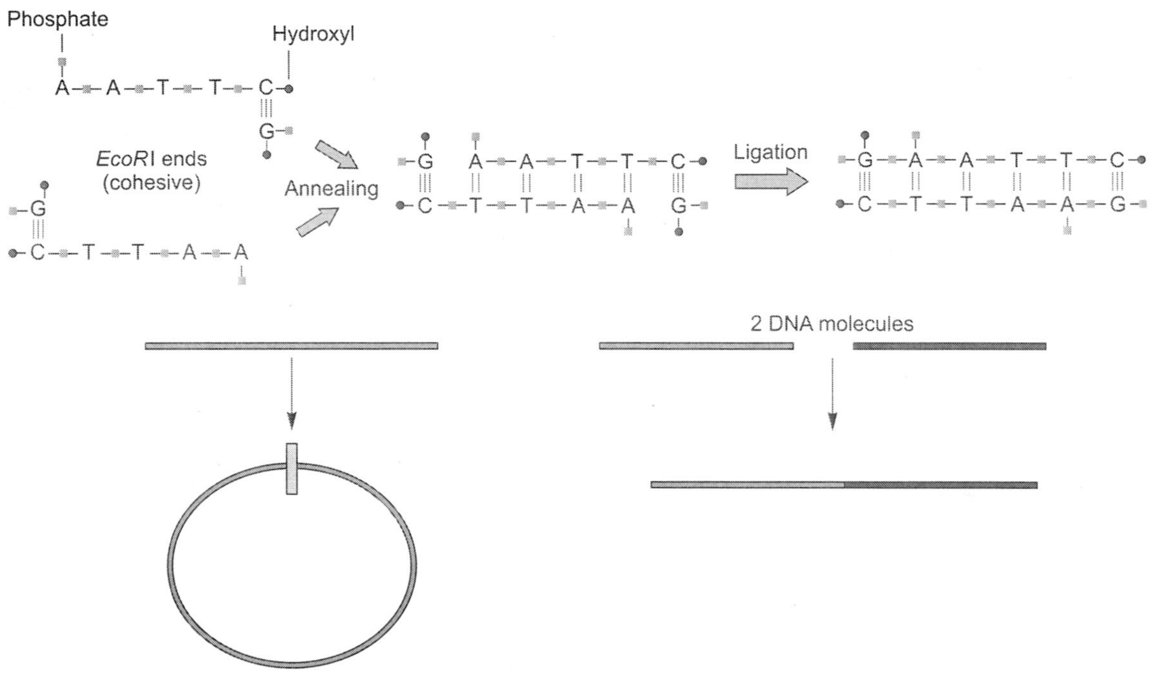

Fig. 19 Ligation reaction: joining of two molecules of DNA using enzyme DNA ligase.

at room temperature for 1–2 hours or 4°C overnight work equally well. Ligase activity is observed at wider range of temperatures for procedures like sub-cloning, but library construction demands high ligation efficiency which can be achieved at optimum temperature of 16°C.

As DNA fragments having sticky ends can be efficiently ligated, different methods are available for changing blunt ends into sticky ends. To accomplish this conversion, linkers and adaptors are developed which can be ligated to blunt ended fragments. These linkers and adaptors are short double stranded molecules possessing a recognition sequence for a restriction enzyme, so that when digested the appropriate enzyme, it produces a sticky end. Another way to create sticky ends involves addition of nucleotides to 3′ end of the fragment by the enzyme deoxynucleotidyl transferase known as **homopolymer tailing.** If the reaction mixture comprises the DNA, enzyme, and only one of the four nucleotides, then only provided nucleotide is incorporated in the new stretch of single-stranded DNA.

HOMOPOLYMER TAILING

In this technique, a nucleic acid homopolymer (a long chain of the same nucleotide over and over) is attached to the end of a piece of DNA as a part of DNA cloning. This technique is especially useful in cDNA cloning to get the cDNA inserted into the cloning vector. A nucleic acid homopolymer which contains the complementary nucleotide to the cDNA tail is attached to the vector, which then pairs with the tail on the cDNA (Fig. 20).

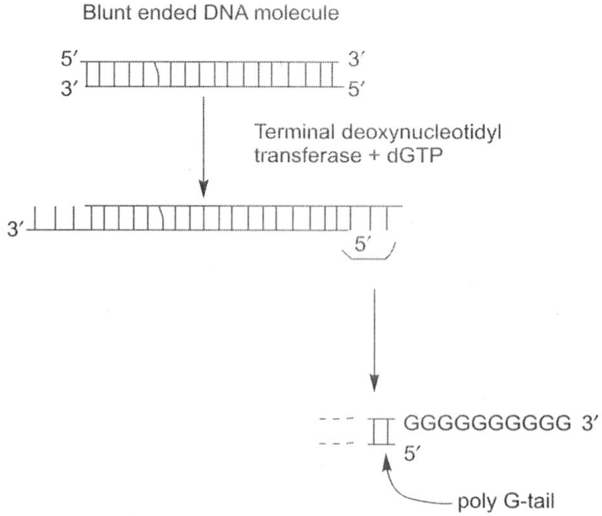

Fig. 20 Homopolymer tailing showing addition of dGTP at 3′ end by terminal deoxynucleotidyl transferase.

In addition to the above discussed enzymes used in gene manipulation, Table 4 also enlists some of the common enzymes that are used in recombinant DNA technology.

Table 4 Some of the Enzymes that are Used for Recombinant DNA Technology

Enzyme	*Activity*
Alkaline Phosphatase	Removes 5′ phosphate groups of DNA molecules; bacterial alkaline phosphatase (BAP) is more stable, but less active than calf intestinal alkaline phosphatase (CIP).
DNase I	Degrades double-stranded DNA by hydrolyzing internal phosphodiester linkage.
E. coli exonuclease III	Sequentially removes nucleotides from 3′-OH ends of DNA molecules except from protruding 3′- OH termini.
Klenow fragment	Proteolytic product of *E. coli* DNA polymerase I that has both polymerase and 3′- exonuclease activities and no 5′- exonuclease activity because fractionation of the digestion products removes the fragment with the 5′- exonuclease activity' a Klenow fragment with only DNA polymerase activity due to mutation in the 3′- exonuclease sequence.
Mung bean nuclease	Single-stranded DNA and RNA endonuclease.
Nuclease Bal-31	Degrades both 3′ and 5′ ends of DNA without internal cleavages.
Poly (A) polymerase	Add AMP from ATP to the 3′ end of mRNA.
Reverse transcriptase	Retroviral RNA- directed DNA polymerase.
RNase H	Degrades the RNA strand of a DNA: RNA hybrid molecule.
S1 nuclease	Degrades single-stranded DNA.
T4 polynucleotide	Catalyzes the transfer of the terminal (γ) phosphate from a kinase nucleoside 5′- triphosphate to a 5′- hydroxyl group polynucleotide.
T7 DNA polymerase	DNA polymerase and 3′- exonuclease activities.
Taq DNA polymerase	Heat-stable DNA polymerase from Thermus aquaticus used in the in vitro amplification of DNA.
β-Agarase I	Digests agarose; used to retrieve separated DNA molecules from agarose gels.

To summarize, availability of these primary tools, i.e., transformation method, cloning vectors and restriction endonucleases and ligases have made all that possible which at one time looked impossible for biologists and if we have these tools available, we can easily start experiments in the *r*-DNA technology.

The importance of transformation methods, cloning vectors and number of enzymes especially restriction endonucleases and other enzymes are demonstrated. The DNA to be cloned is cleaved by restriction digestion, the cloning vector is also digested with the same restriction enzyme and two DNA molecules are ligated and then the ligation mix is transformed into *E. coli* host cells and the recombinant colonies are then selected under appropriate antibiotic pressure (or any other selectable trait; Fig. 21).

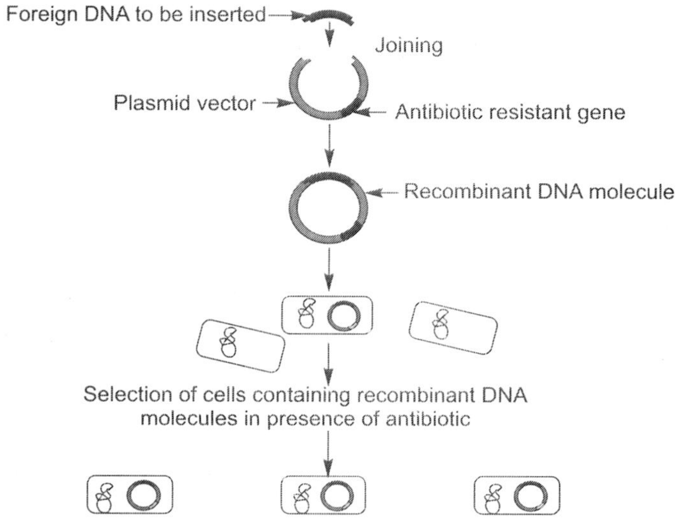

Fig. 21 Construction of a recombinant DNA molecule.

As mentioned earlier, these primary tools were eventually supplemented by many other new methods including Polymerase Chain Reaction (PCR), Southern Blot Hybridization (SBH), automated DNA sequencing and automated, easy and rapid methods for isolation, cloning and characterization of genes.

These methods have become very important in molecular biology and have been covered during various protocols described in this manual.

As a result molecular biology has found an application in almost every discipline of biological sciences. Some of these new tools are also described with this background. Let us now move on to the third aspect, i.e., application of biotechnology. But before that we will redefine biotechnology in the light of what already has been mentioned. Molecular biotechnology is a discipline which arose as a result of fusion of research in several disciplines particularly in microbiology and biotechnology in the late 1970s.

DNA SEQUENCING

DNA sequencing involves various biochemical methods used to determine the order of nucleotide bases (adenine, guanine, cytosine, thymine) within a sequence of DNA oligonucleotide. Within this method of automated DNA sequencing the dideoxynucleotides are fluorescently labeled with dyes fluorescing at different wavelengths. As these are incorporated into the DNA sequence, the wavelength at which they fluoresce is captured by a CCD (charged coupled device) camera and saved. Analysis of this sequence of wavelengths, in turn, results in inference of the DNA sequence. Modern automated sequencers used nowadays are very efficient and give reduced run up time also. DNA separation in this case is done by capillary electrophoresis, detection and recording of dye fluoresce and the data output is in the form of fluorescent peak trace chromatogram. High throughput sequencing methods are also in the making to cut down on the cost of sequencing and for better efficiency (Table 5).

Generation of the raw data, i.e., DNA sequencing is the maiden step in any genome project. Sanger's method of DNA sequencing based on the dideoxy chain termination was the leading platform used for sequencing of the first bacterial genome, and it has been the mainstay of many molecular biology laboratories for the last 30 years. Being the primary platform, it is designated to be the first generation sequencing technology. While on one hand it is simple and gives a read length of around 300-400bp with high accuracy, it has its own demerits in being very costly and time consuming as it requires construction of libraries too. But since this method works on a small scale as well as a large, one with minimum artifacts, it is still in use.

Sanger's method has been gradually replaced by next generation sequencing (NGS) technologies over the last few years as these NGS platforms allow exceptionally rapid generation of high throughput data. These include the second generation sequencing technologies like pyrosequencing, AB SOLiD, Illumina and many more. NGS generates short reads sequence data, but in huge amounts as compared to Sanger's long reads about 300–400 bp. The two best performing platforms known so far, i.e., AB SOLiD and Illumina generate 35–75 base pair reads and are very cheap, while pyrosequencing which comparatively is a lower-throughput platform can manage 400 base pairs in an intermediate cost range. Therefore, major concern with all the sequencing technologies is just the read length.

However, "end sequencing" to complete genome sequence and other small scale experiments are still performed using dideoxy method of sequencing as these are not easily handled by NGS technologies.

Table 5 Comparison between Different Sequencing Techniques on the Basis of their Run Time, Read Length, Accuracy and Cost

	Roche/454	*Illumina/ SOLEXA*	*ABI SOLiD*	*Polonator*	*Heliscope*	*Ion Torrent*
Method for sequencing	Polymerase (pyrosequencing)	Polymerase (reversible terminator)	Ligase (octomer)	Ligase (nonamer)	Polymerase (single molecule)	Polymerase (single molecule)
Template amplification method	Emulsion PCR	Bridge PCR	Emulsion PCR	Emulsion PCR	None (Single molecule)	None (single molecule)
Throughput	400-600 MB	17 GB	10–15 GB	~10 GB	21–28 GB	100 MB
Run time	10 h	4 days	7 days	4 days	8 days	1–2 h
Read length	~400 bp	~40 bp	~50 bp	~15 bp	~30 bp	100–200 bp
Accuracy	99.74%	99.99%	99.7%	98%	99%	>99.995%
Cost	$60.0/MB	$2.0/MB	$2.0/MB	$1.0/MB	$1.0/MB	$ 500/run

MICROARRAY

Microarray is used as a tool for analysing gene expression that consists of small membrane or glass containing samples of many genes arranged in a regular manner. The whole process is based on hybridization probing, a technique that uses fluorescently labeled nucleic acid molecule as "mobile probe" to identify complementary molecule, sequence that are able to base pair with another (Fig. 22).

COMMERCIAL IMPORTANCE OF BIOTECHNOLOGY

Biotechnology has become so common and its impact is so high that even in general public we talk about it. Biotechnology is talked about in movies like Jurassic Park with a plot to clone dinosaur. Molecular biotechnology is the fusion of researches primarily in biochemistry, genetics, chemical engineering and cell biology. But major contribution has come from molecular biology and microbiology. Biotechnology is finding applications in solving environmental problems, crop production, newer drugs and vaccines, in diagnostics and has contributed simultaneously to the development of gene therapy. We read in newspapers about commercialization of biotech and release of engineered products in the market. We also come across several topics written by the critics of biotechnology about the possibility of dire consequences of genetic engineering. Numbers of recombinants or engineered

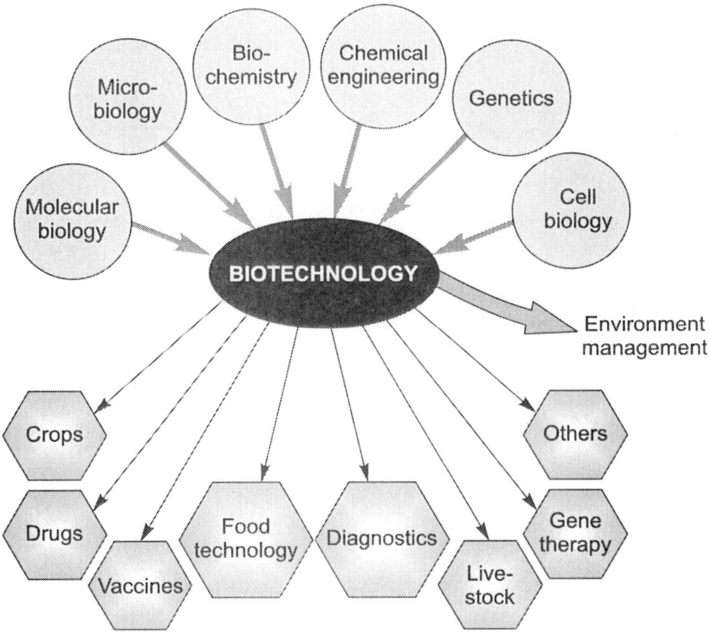

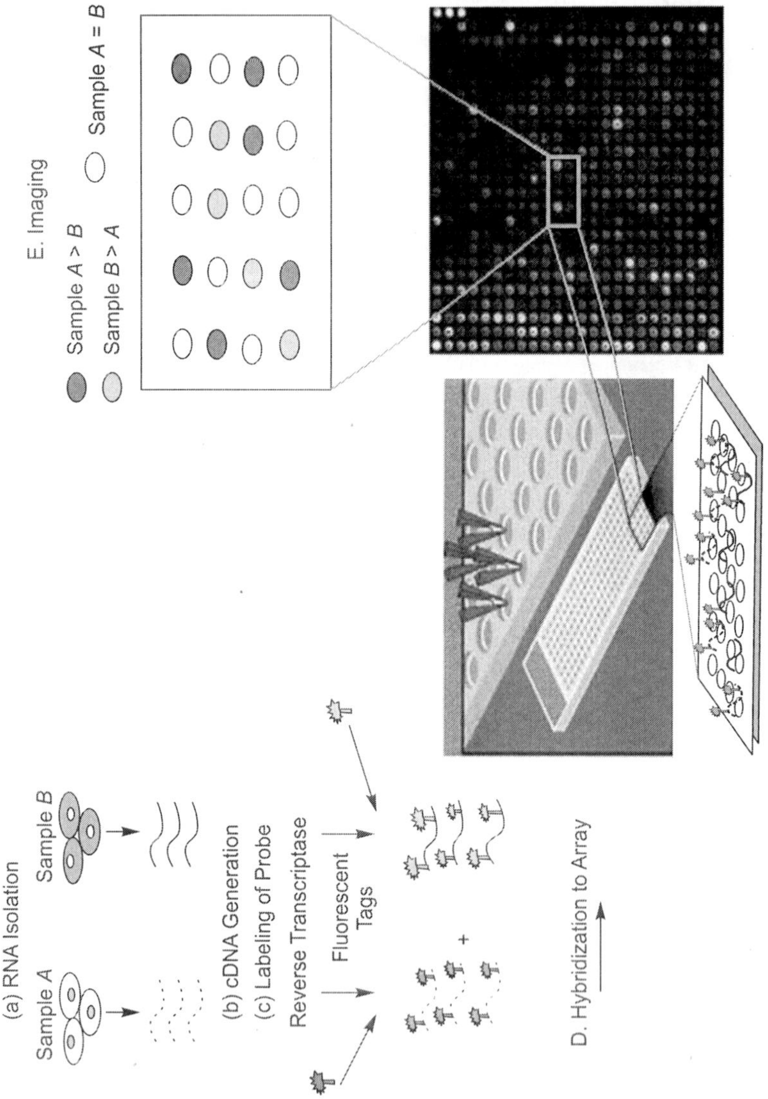

Fig. 22 Schematic representation of the protocol followed during microarray analysis.
Source: http://www.eufic.org/upl/1/default/img/Fig%203%20Microarray.jpg

molecules are available like insulin, growth hormone, hepatitis B, coagulation factor, interleukin, interferons, etc. (Table 6). Commercialization and release of these have marked a revolutionary change for the betterment of human race.

Table 6 Recombinant Molecules and their Applications

Product name	*Protein type*	*Application*	*Company*
Adagen (Adenosine deaminase)	An enzyme	Severe combined immunodeficiency disease (SCID)	Enjon
Genotropin (Recombinant growth hormone)	A hormone	Growth hormone deficiency (GHD) in children	Pharmacia & Upjohn
Humalog (Recombinant human insulin)	A hormone	Diabetes	Eli Lilly
Nabi-HB (Anti-Hepatiti B)	An antibody	Hepatitis–B	Nabi
Nova Sevwn (Recombinant coagulatgion factor VIIa)	A modified factor	Hemophilla patients with inhibitors	Nova Nordisk
Ontak (Diphtheria toxin-interleukin-2)	A fusion protein	Cutaneous T-cell lymphoma (CTCL)	Ligand Pharmaceuticals
Roferon-A (Recombinant interferon alfa-2a)	A modifier	Hairy cell leukemia or AIDS-Related Kaposi's sarcoma	Hoffmann-La Roche

Several valuable products have been produced as a result of the use of molecular biology. Some of them have been listed and list is growing and new molecules are being added to it. These are adenosine deaminase, growth hormone, insulin, hepatitis B and interferon Alfa-29. The availability of human insulin was a good news for millions of patients suffering from diabetes.

Recombinant Insulin

The first licensed drug produced by genetic engineering was insulin. Traditionally, it is extracted from pancreatic cells of pigs and cows. There are several limitations of this insulin, but recombinant insulin appears to have solved this problem to some extent. The initial approach was to construct synthetic genes from oligonucleotides that separately encoded

A and B chains of the protein. They were individually inserted into *E. coli* gene β-galactosidase and thus, the bacterium produced the β gal insulin fusion proteins.

Treatment with cyanogen bromide clipped off the A chain with 21 amino acids and B chain with 30 amino acids (51 amino acids) from the fusion protein. The approach has been refined and recombinant insulin is produced under the commercial name Humulin (Fig. 23).

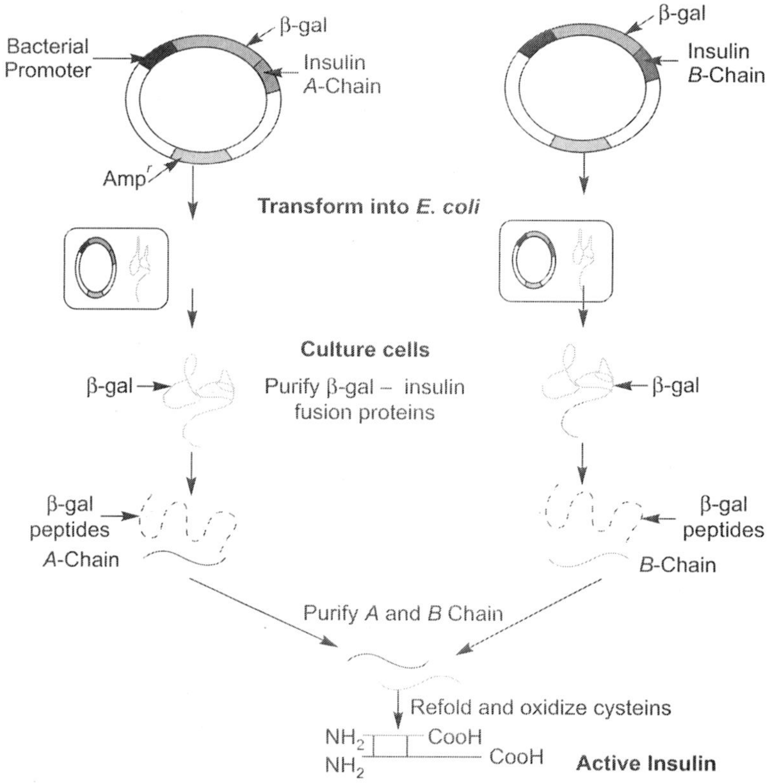

Fig. 23 Recombinant insulin and its expression in *E. coli* cells.

Growth Hormone

Growth hormone is a protein of 191 amino acids and is being used to treat children born with growth hormone deficiency. Its cDNA was prepared and cloned and initial sequence for 1–24 amino acids was replaced with synthetic oligonucleotides without changing the sequence of amino acids. The fragment was cloned in *E. coli* expression vector and growth hormone accumulated in the cell. In order to get it excised to the periplasmic space, the mammalian signal sequences were replaced by bacterial signal sequences and thus the recombinant growth hormone was secreted into the periplasmic space (Fig. 24). R- HGH is manufactured by Novo nordisc of USA and it is sold in India by Med India Limited situated at Bangalore. But the treatment is very expensive.

Similarly, the availability of recombinant growth hormone is a great benefit to the children born with hypopitutarism. Other valuable products now being made by genetic engineering are proteins used in healing burns, hemophilia, cancer, vaccines and the development of diagnostic tests.

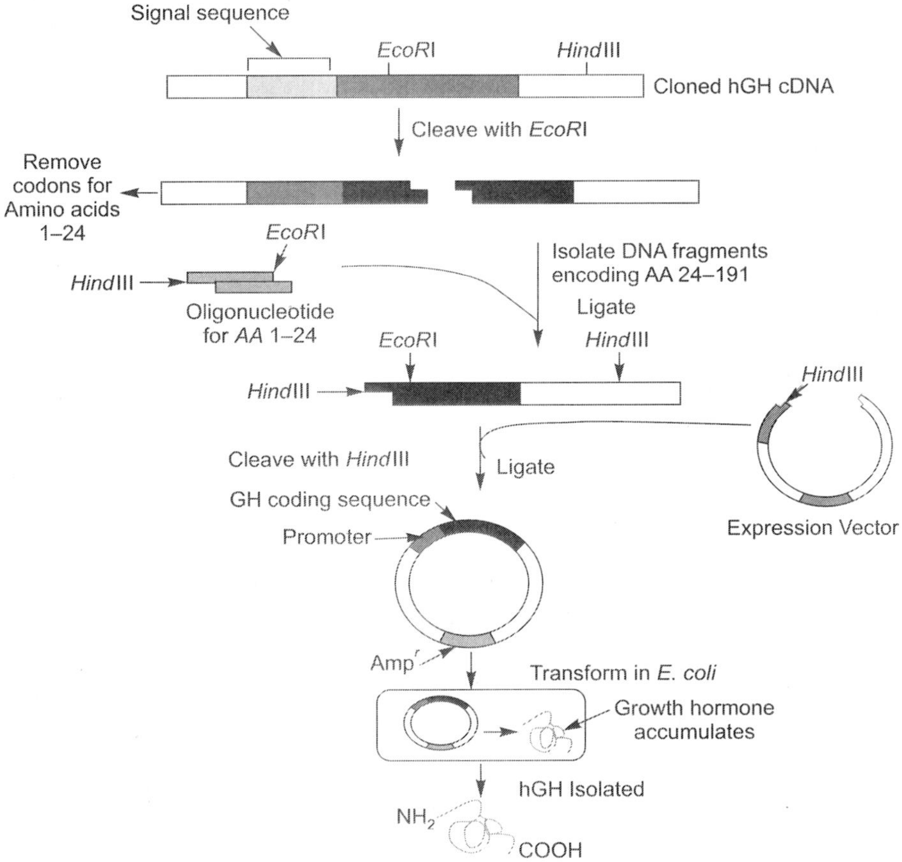

Fig. 24 Bacterial production of recombinant human growth hormone.

Creation of Superbug

Superbug was created by A.M. Chakrabarty and it can degrade petroleum spills. He selected four strains of *Pseudomonas* which could degrade camphor, octane, xylene and naphthalene. The degradation of these hydrocarbons was mediated through catabolic genes on the plasmids.

Mating of strain A and B led to the formation of a recombinant strain E which could degrade both camphor and octane, while mating of strains C and D resulted in the formation of strain F containing two plasmids with ability to degrade xylene and naphthalene. Mating between E and F led to the development of a strain G which could degrade all the components generally found in petroleum (Fig. 25). Although this strain has not been used for clean-up of oil spills, it has played a critical role in the development of biotechnology industry. This superbug contains the plasmids from different *Pseudomonas* strains that can degrade various organic compounds together. The inventor of this superbug A.M. Chakrabarty was granted US patent in 1980. This was the first historical decision in which a court permitted a living organism or recombinant bacterium to be patented by the scientists.

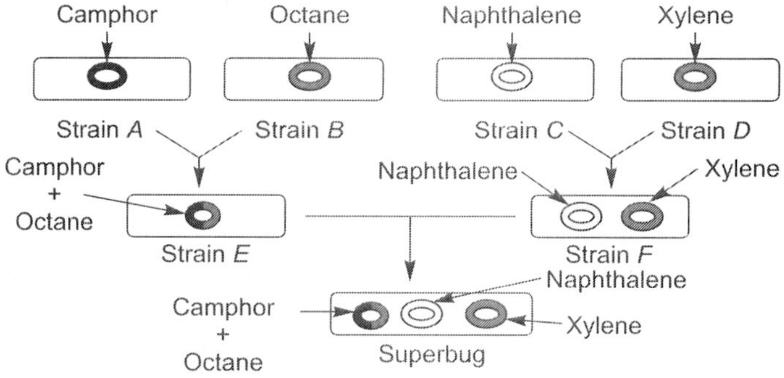

Fig. 25 Schematic representation of process of construction of superbug.

The degradation of most of the aromatic hydrocarbons like benzoate, xylene, toluene and HCH is mediated through catabolic genes generally located on the plasmids and these plasmids have been named as TOL plasmids. The degradation of these compounds is mediated through an intermediate catechol which is then converted into intermediates of Krebs cycle. The enzymes and the catabolic pathway have been completely identified and studied. The catabolic genes XYZLEGFJKIHS of the upper pathway are activated through a promoter Pm which is, in turn, activated by a complex formed by XYLS + benzoate or 4-methyl benzoate, but the industrial effluents in addition to benzoate or 4-methyl benzoate also contain 4-ethyl benzoate which cannot induce promoter Pm alone. But once Pm is induced, the benzoate can degrade even 4-ethyl benzoate to 4-ethyl catechol. But this intermediate is again toxic to catechol 2,3–dioxygenase. To overcome these two problems by applying classical methods *Pseudomonas* strains were generated which could accept 4-ethyl benzoate as an inducer and now 4-ethyl catechol was not toxic to the important enzyme, 2,3-dioxygenase.

In an another case, toluene deoxygenase which is responsible for the degradation of toluene in *P. putida* was also found to degrade trichloroethylene (which is used as a solvent and a degreasing agent and common contaminant of groundwater and industry). But the intermediate formed in the process was still more dangerous and the degradation was much slower. This problem was overcome by cloning TOL deoxygenase genes along with TOL promoter of *E. coli* which could be induced by the action of IPTG. As a result of not only the degradation process was enhanced, but the TCE was completely mineralized. *P. putida* also contains plasmids responsible for the degradation of naphthalene and in one experiment when scientists were analyzing the genes present on the plasmids they found out that *E. coli* clone turns the entire medium blue. The analyses revealed that this clone contained a fragment of DNA which encoded naphthalene deoxygenase and tryptophanase of *E. coli*. Converted tryptophan to indole which was acted upon by naphthalene spontaneous dehydration resulted in the formation of Indoxy which upon oxidation was converted to indigo.

Traditionally, indigo is an important blue pigment which is used to dye both cotton and wool and originally isolated from plants and currently produced through chemical synthesis. The availability of molecular techniques has made it possible to rewrite and reconstruct the phylogenetic tree.

Transgenic Plants

Exciting results of genetic engineering in higher organisms are also being reported by plant molecular biologists. Plants enjoy the special status of being able to regenerate from single cell. Another advantage was the availability of a natural system, a Ti-plasmid of *Agrobactrium* which contained T-DNA. *Agrobactrium tumefaciens* is a pathogen that causes tumors called crown galls. The tumors are induced by the plasmid which is called the Ti plasmid and it contains T-DNA region that integrates into the plant genome during infection process (Fig. 26A). This property has been exploited by plant molecular biologists to generate transgenic plants. Similarly, luciferase gene has been cloned into Ti- plasmid and transfected into plant cells which then regenerated into complete plant and the plant glows during night (Fig. 26B).

Initially, it all started from tobacco and potato, but now the technology has been perfected for almost all types of plants. Even there are hopes of using plants as bioreactor for generating important molecules, for example, human interferon gene has already been cloned and expressed in plants.

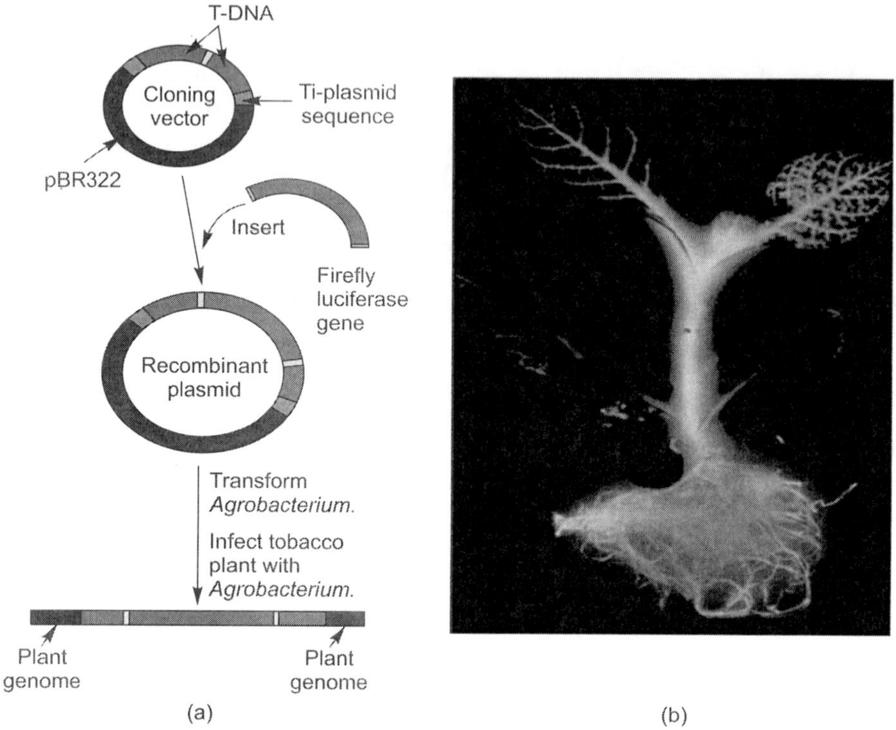

Fig. 26 A. Generation of recombinant plasmid by insertion of firefly luciferase gene to the Ti-plasmid of *Agrobacterium tumiefaciens*. B. Transgenic plant expressing the firefly's luciferase gene upon transformation by *A. tumiefaciens*.
Source: http://www.stc-orlando.org/education/highsch/winning_entries/2005_JFromm.pdf

Bt cotton

One of the very important and also commercial applications is the Bt-cotton. We know transgenic Bt-cotton contains a gene from *Bacillus thuringinesis* that produces endotoxin which has insecticidal properties. This gene was cloned in plant cells and transgenic plants were raised. Use of Bt-cotton has already been started in a number of countries including India.

Edible vaccines - A Biopharming Dream

Edible vaccines can be generated by first cloning the pathogen protein gene and then inserting the cloned gene into the DNA of plants like potato, banana, tomato, etc. Humans can eat fruits of the transformed plants. The body produces antibodies against the pathogen protein and the individual is immunized against the pathogen.

Vaccines for diarrhea, hepatitis B and measles are generated by this method.

Terminator technology

Termination technology is the name given to a method which is meant for restricting the use of genetically modified plants by causing second generation seeds to be sterile. The technology has not been commercialized for stakeholders are concerned that it might lead to dependence of poor farmers on the technology, although it was developed by U.S. Department of Agriculture and Delta and Pine Land Company in 1990s.

The technology is a form of **Genetic Restriction Technology** (GURT), which further is of two types:

1. **V-GURT:** This produces sterile seeds, so that the seeds from the current crop cannot be used by farmers for future. The farmers would again use hybrid seeds from seed production companies. The term V-GURT arises from the fact that the technology is restricted at plant variety level.
2. **T-GURT:** This kind of crop does not express the genetic enhancement within itself until it is treated with an inducing chemical sold by the company. Farmers would be able to save their seed for future use, but will have to purchase the activator compound. The technology is restricted at trait level and hence springs the term T-GURT.

The possible advantages of using seed terminator technology span from an incentive to development of new plant varieties, improved farm management, improved grain quality as well as biosafety. But there exists various disadvantages as well, due to which commercialization of this technology has been banned. The V-GURT plants could cross-pollinate and the trait could get transferred to wild types and cause ecological problems. Concerns are also present for safety of food produced, and economic concern of making small farmers as well as entire rural community dependent on agro industry corporation seeds.

Scientists have now cloned a cassette of three genes which contains a toxin, a recombinase and a repressor gene. Repressor gene produces a repressor protein that, in turn, binds to the DNA sequence between recombinase gene and the promoter sequence, thus halting the expression of recombinase and the production of recombinase protein does not occur. If the spacer between the promoter sequence and toxin gene is not cut then no toxin is produced. But if inducer is applied to the seeds no repressor blocks the production of recombinase and thus the production of toxin and development of embryos (Fig. 27).

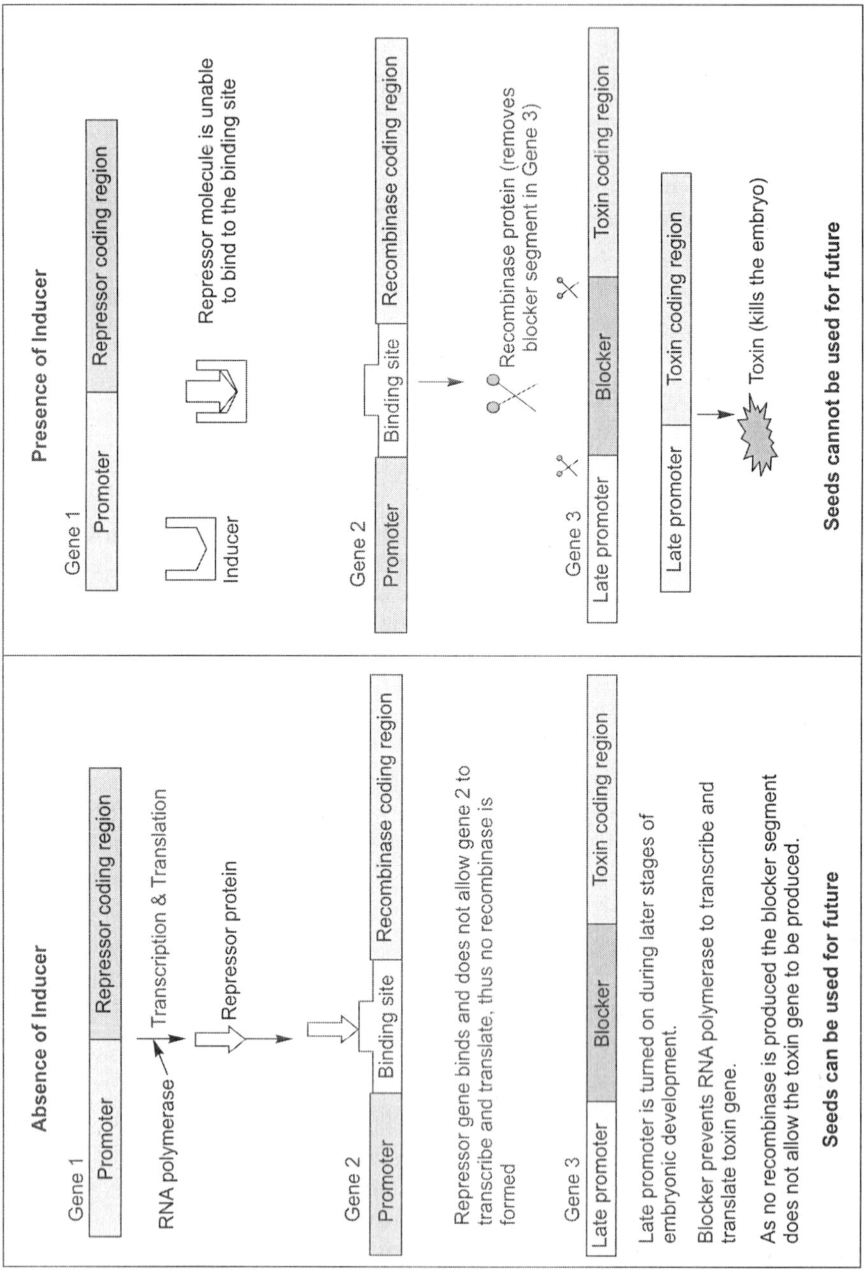

Fig. 27 Schematic representation of the seed terminator technology.

Gene Therapy

Another area which is going to be affected by biotechnology is 'gene therapy' which means introduction of DNA or RNA into the cells with the intention of clinical benefits. The candidate diseases for which field trials are already at advanced stages are cancer, oncogenic inherited disorder, HIV and cardiovascular infections. The first child patient Ashanti Desilva (Fig. 28), whose treatment was approved by the US government, was suffering from severe combined immune deficiency syndrome (SCID). The defect is due to the deficiency of the enzyme — adenosine deaminase (ADA), which is inherited from each parent. The absence of this gene or enzyme impairs the normal functioning of immune system.

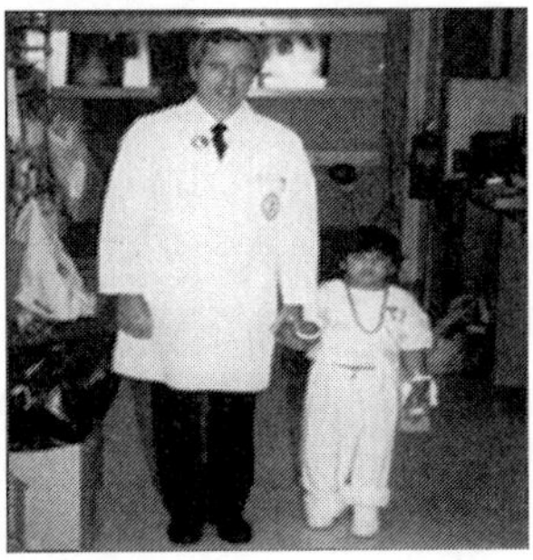

- The first recipient of federally approved gene therapy on September 14, 1990
- Suffering from Severe Combined Immunodeficiency caused by ADA (Adenosine deaminase)

Dr. W. French Anderson and his gene therapy patient Ashanti DeSilva in 1990

Fig. 28 The miracle called gene therapy

The treatment Ashanti Desilva received was removal of white blood cells from her body. These blood cells were rectified for deficient gene and corrected cells were then again transfused (Fig. 29). She is getting regular infusion of such cells and leading a healthy normal life. Hundreds of methods have now been approved and field trials are being conducted.

Animal Cloning

In 1997, at the Rosalin Institute in Scotland, scientists reported the cloning of Dolly the sheep who became the world's first adult mammal to be cloned. Wilmut and his colleagues took somatic cells from the udder of an adult sheep which already contained diploid number of chromosomes. These cells were fused with enucleated cell of another sheep by subjecting them to electrical pulses. The fused cells were cultured and conditioned. The embryos raised in his way were then transplanted into the womb

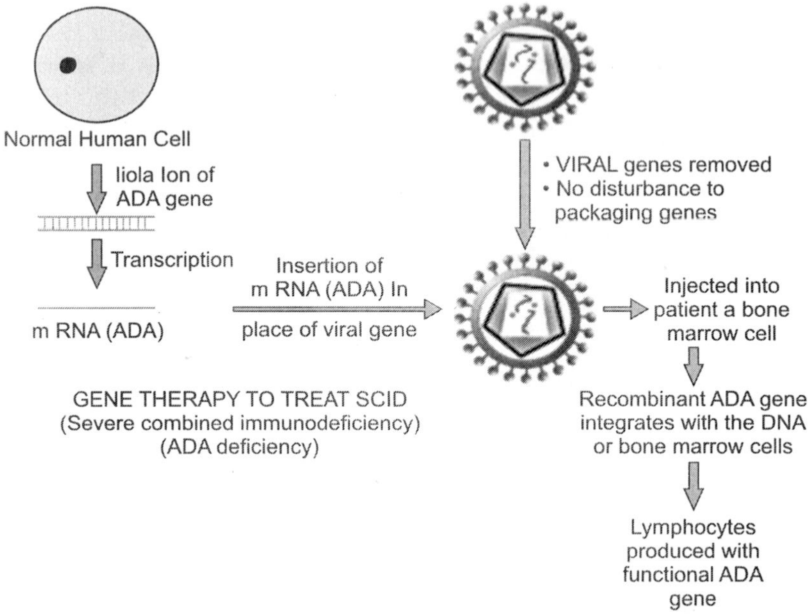

Fig. 29 An outline of application of gene therapy for treatment of SCID.

of a surrogate sheep (third sheep). Five months later, a sheep was born named Dolly – a clone of her mother and without a biological father (Fig. 30). The cloning of Dolly has raised much interest and debate over the potential of cloning mammals including humans.

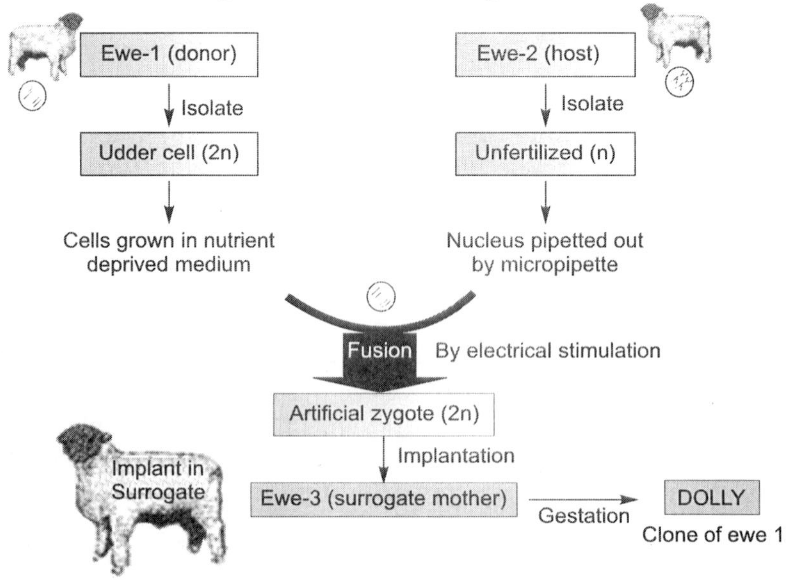

Fig. 30 Schematic representation of cloning of Dolly in the year 1996.

After cloning of Dolly, number of other animals goats, buffalos, monkeys have been cloned. There are unconfirmed reports of cloning human beings and already serious debate is going on this issue the world over. One of the immediate effects of cloning is developing pigs as donor for organ transplantation. With development of molecular techniques hundreds of women were treated for infertility and have had their embryos screened for genetic defects. The controversial technique called pre- implantation genetic diagnosis involves selecting the embryos, that are then checked for defects and only those free from defects are implanted in the womb. This has been tried in UK on women who had family history of genetic disorders like cystic fibrosis, hemophilia and muscular dystrophy.

Stem Cell Technology and Its Applications

Stems cells found in all multicellular organisms have the ability to renew themselves by mitotic cell division and are able to differentiate into different types of specialized cells. Embryonic stem cells and adult stem cells form the two main broad categories of the stem cells. Since stem cells possess the ability to be transformed into any specialized kind of cell, such as the muscle or the nerve through cell cultures, their proposed applications in medical therapies are immense.

Few applications are listed below:

1. Transplantation Research

Shortage of donor organs makes it unlikely that a damaged and diseased organ gets transplanted. But here, stem cells can act as savior. Stem cells are now being used to regenerate an entire functional organ, which can be derived from the suffering patient itself and thus can be transplanted into the patient without any fear of immune rejection.

2. Basic Research Applications

Stem cells can be some day be used to study the fundamental events in the developing embryo. This, in turn, might lead to understanding of birth defects and may lead to approaches to correct them or even prevent them. The genes, molecules, growth factors and the nutrient involved can be studied, so that specialized cells may be produced in the laboratory.

3. Therapeutic Delivery Systems

Genes can be transferred into specific cells or tissues in the body by using stem cells as vehicles of transport. These will find major applications in cancer research. These will deliver treatment directly to the cancerous cells, destroy or modify them without causing any secondary damaging effects as in the case of radiotherapy and chemotherapy. Somatic Cell

Nuclear Transfer (SCNT) is used as a procedure in therapeutic cloning. Cells in the human body other than egg and sperm are called somatic cells. In this procedure, a nucleus from somatic cell of the body is extracted and injected into the enucleated egg (nucleus from the egg cell is removed previously). Therefore, the egg has new genetic material from somatic cell nucleus which stimulates the egg to divide and forms blastocyst having two layers of cells; an outer and inner layer. Clusters of cells forming inner cell mass rich in stem cells are harvested and used to form embryonic stem cell lines, which can be infused into patient's tissues to provide the required structure and functions in the affected patient (Fig. 31).

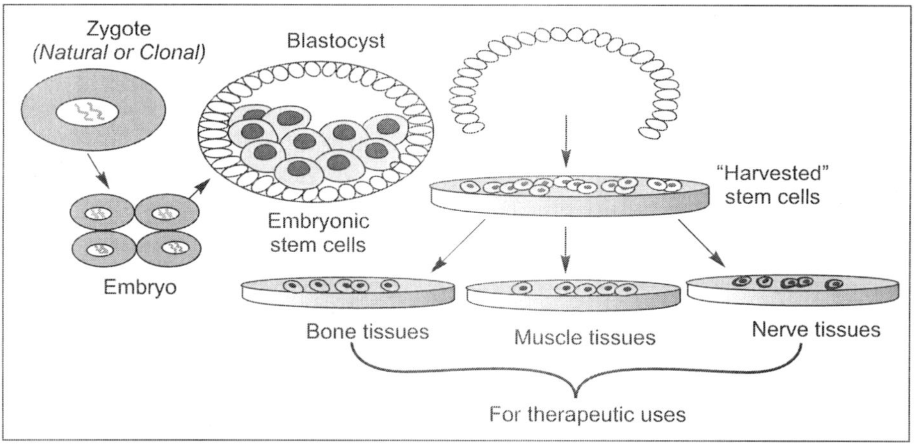

Fig. 31 Schematic representation of the process of therapeutic cloning. A nucleus from somatic cell of the body is extracted and injected into the enucleated egg (nucleus from the egg cell is removed previously). Therefore, the egg has new genetic material from somatic cell nucleus which stimulates the egg to divide and forms blastocyst having two layers of cells; an outer and inner layer.

Other applications

Animal models are used in pharmaceutical research for testing of various therapeutics, however their direct effects on human cells cannot be predicted always, and candidate drug trials on human stem cells can be substantially beneficial in future. Also, detoxification competency can be studied by using stem cell technology for development of specialized liver cells which can help in early detection and prevention of adverse side effects in patients. Linking stem cells with the data obtained from human genome project can likely have many unforeseen benefits.

Scientists have cloned pigs which can be now good donors for human organ transplantation. Generally in pigs, the endothelial cells produce a sugar which is recognized as foreign body by the human system and thus organ transplant is rejected. Scientists have recently cloned pigs in which glactosyl transferase gene was mutated in the stem cell and then pigs so produced lack the sugar and thus can become suitable donors. The future of gene therapy lies on the stem cell technology.

For certain diseases, traceable to single defective gene, such as cystic fibrosis and other diseases affecting human immune system, scientists are already carrying out human gene therapy on experimental basis. The goal of such therapy is to transplant a functioning gene into the cells that will then manufacture the normal gene product in amounts large enough to alleviate the condition. Diseases affecting blood and bone marrow cells, which include stem cells that give rise to blood cells, are the prime candidates. Most of the gene therapy trials in progress are yielding promising results suggesting that genetic engineering will eventually provide means for permanent correcting, rather than just treating many genetic diseases. The ultimate cure, of course, would be to replace malfunctioned alleles from germline—sperm and eggs. But it raises serious ethical and social questions.

Introduction to Gene Manipulation

Anjali Saxena & Ankita Dua
E-mail: *anjali.saxena1403@gmail.com, ankitadua1987@gmail.com*

Gene manipulation refers to various molecular biology techniques involved in the creation of recombinant DNA. "Manipulating the gene" which is the unit of "heredity" involves basic techniques that are of growing interest and importance in the biotechnology world. Techniques start from basic isolation of DNA/RNA to its transfer to a foreign host cell and its further manipulation.

Genes of interest that need to be targeted into the host organism involve the use of restriction enzymes, ligases, etc., in order to cut or join the fragment of DNA and the success of such experiments is evaluated by visualizing the molecules using gel electrophoresis. Amplification of a gene fragment can be achieved by polymerase chain reaction and this yields ample material that can be proceeded with for sequencing.

Foreign DNA is inserted into host cells via plasmid vectors as constructs. In order to express a gene into a host cell, mostly the constructs consist of promoter and terminator to initiate and end the transcription, respectively. Besides, there are also selectable markers that should confer antibiotic resistance, as this is required in order to determine which cells are transformed with the new hybrid gene. About 1% of bacteria are naturally able to take up foreign DNA, but it can also be transformed into bacteria by means of electroporation or heat shock transformation.

Further confirmation that the organism contains a new gene is done using PCR, southern hybridization or sequencing.

This book further entails the techniques involved in gene manipulation that are the grass roots level experiments done in a molecular biology laboratory. The protocols have been enlisted with precision along with the dos and don'ts for each.

Media and Chemical Preparation: This chapter deals with the general chemicals and media composition which are routinely being used for culturing the Gram-positive and Gram-negative bacteria.

Growth and maintenance of bacterial cultures: In molecular biology many microorganisms are involved in the experiment and we need specialized techniques to handle them. It includes methods of streaking, culturing, autoclaving, etc.,

Isolation of genomic DNA: Isolation of genomic DNA from cells is an important protocol used in molecular biology. It involves: harvesting of cells, disruption of cell membrane, phenol chloroform treatment and precipitation of DNA.

Isolation of plasmid DNA: Plasmids are double stranded, extra-chromosomal DNA of bacterium. They are important tools in molecular biology and in biotechnology they are commonly used for the expression of genes. For the isolation of plasmid DNA, bacterial cell harboring it are first grown in the presence of antibiotic pressure selective for plasmid. It is then followed by lysis of cells cleaving phospholipid bilayer of the membrane, precipitation, centrifugation, removal of supernatant and isolation of plasmid.

Isolation of RNA: Isolation of RNA is the first step in various molecular biology techniques like northern blot analysis, transcriptome analysis, cDNA library construction, real-time PCR, etc. RNA being unstable, the sample collection and protection plays an important role in the effective isolation and purification protocol. The isolation involves important steps of bead beating, lysis of cells, silica spin filter wash and elution in RNAase free buffer.

Isolation of Proteins: Protein purification has its importance in various commercial applications like enzymology, pharmaceutical industry, etc. Extraction of proteins from the cells requires proper techniques that are intended to isolate protein of interest from a complex mixture, usually cells, tissues or whole organism. The steps for extraction of proteins include freezing, thawing, sonication, homogenization and filtration by organic solvents.

Agarose gel electrophoresis: It is the easiest technique for separating DNA fragments based on their molecular weight in a matrix of agarose. Nucleic acids are separated by applying electric field that propels the negatively charged DNA molecules to move towards the positive electrode through the agarose matrix.

Pulse field gel electrophoresis (PFGE): This technique allows separation of larger pieces of DNA than conventional gel electrophoresis. Instead of constantly running the current in one direction, in PFGE, it is altered in different directions on regular intervals that ensures separation of the Mb sized DNA molecules.

SDS polyacrylamide gel electrophoresis: It is a technique used in biochemistry and molecular biology for the separation of proteins based on their electrophoretic mobility. For separation of proteins, sodium dodecyl sulfate (SDS), an anionic detergent is used that imparts negative charge to the linearized proteins.

Purification of DNA: Recovery of DNA from agarose gels is an essential step in order to design further experiments like restriction digestion, cloning, etc.

Restriction digestion: It is the process of cutting DNA molecules into various fragments with the help of restriction enzymes. It plays an important role in construction of recombinant DNA molecules as well as to determine number of restriction sites in DNA.

Ligation: DNA ligation is one of the most important tools in genetic engineering and molecular cloning experiments. When a circular strand of DNA is cut with restriction enzymes, then DNA ligase seals the nick and joins the two DNA molecules together.

Polymerase chain reaction (PCR) and primer designing: PCR technology has potential applications in DNA fingerprinting, forensics, predicting phylogeny, molecular cloning, sequencing, etc. PCR is a method to produce millions of copies of DNA (gene, gene fragment, etc.) from a mixture of total DNA in a couple of hours. Primers are the starting raw materials used to initiate DNA amplification reactions and their sequence is crucial to the process.

Cloning and cloning vectors: DNA cloning is a technique that has revolutionized research at the molecular level and directs assembly of recombinant DNA molecules and their uptake by host cells. Cloning vectors are the simple vehicles that harbor the bacterial resistance gene, origin of replication and multiple cloning site. They are usually optimized for initial cloning of a DNA fragment.

Transformation: Bacterial transformation is the cornerstone of the recombinant DNA technology. It provides in the hand of the experimenter, tools to introduce new genes modify the genomic make up of the organism and thus aids in understanding bacterial genetics. Transformation aids in cloning, bringing genes from two different species together into one. Thus, it finds application in industry and the clinical realm as well. The protocol requires availability of cloning vectors, gene of interest (in plasmid), competent cells (bacterial cells which can take up foreign DNA).

Southern hybridization: It is a technique used for detection of gene of interest in DNA samples. The method combines use of restriction enzymes to cut the DNA molecule, separating the DNA fragments by gel electrophoresis, transferring the fragments to filter membrane by depurination, denaturation in alkaline environment, buffer transfer by capillary action and final transfer of DNA fragments to the nylon membrane. Once the blot is prepared, the detection is done using prehybridization, followed by hybridization to a radiolabelled or non-radiolabelled probe. After hybridization, excess probe is washed away and pattern of hybridization is visualized on X-ray film by autoradiography.

Sequencing: To have a full grasp on manipulating the gene(s) of interest in bacteria, it is important to know what our target sequence is and what we desire to manipulate. This can only be known through DNA sequencing. Although, a wide array of next generation sequencing platforms are available. Sanger sequencing is still the most popular method adopted for regular sequencing of genes.

1
Preparation of Media and Chemicals

Amit Kumar Singh & Pooja Rani
E-mail: *amit.k.singhbhu@gmail.com*
poojaasrivastav@gmail.com

INTRODUCTION

Growth media are usually liquid or gel intended to support the growth of microorganisms. They supply nutrient and essential elements normally required for the growth of microorganisms. On the other hand they are also involved in the isolation and maintenance of pure cultures of bacteria. Further, they are employed in the identification of bacteria according to their biochemical and physiological properties. Our knowledge to study different types of bacteria depends upon their nutritional requirements. However, the growth and continuity of microorganisms depends on favorable growth environment and conditions. For a successful culture of a given microorganism, it is necessary to understand its nutritional requirements and proportions of different constituents of media. The bacteria which are commonly present in the environment are mostly generalist because of their natural capabilities to metabolize a wide array of nutrient sources.

According to the requirement and stability, three types of media are commonly used:

(a) Solid media
(b) Liquid media
(c) Semi-solid media

The manner in which bacteria are cultivated depends on their growth requirements and it varies to different degrees. The major difference among them is that solid and semi-solid media contain a solidifying or gelling agent (such as agar, gelatine), whereas a liquid medium does not.

- **Solid media** contains agar in different proportions, but generally 1.5 or 2 per cent is used. Agar is commonly used for media preparation because of its remarkable physical properties, i.e., it melts at 100°C and remains solid at 40°C. Further, it is not utilized by a large number of bacteria therefore as a component of media is chemically inert. Colony morphology, pigmentation, haemolysis can be examined using solid media. Example: LB agar, nutrient agar, blood agar, etc.

- **Liquid media** contains no agar. It is used for inoculation purpose (primary and secondary inoculation), growth of pure batch culture and for the isolation of pathogens from a mixture.
- Example: LB broth, etc.
- **Semi-solid media** contains 0.5 per cent agar. It is used to check the motility of bacteria.

Example: sulphide-indole-motility and motility test agar, etc.

According to the function, culture media are categorized into 5 types:

(a) All purpose media
(b) Selective media
(c) Minimal media
(d) Differential media
(e) Enrichment media

All purpose media, such as tryptic soy agar (TSA), which supports the growth of most bacteria cultured in the laboratory. They do not contain any special additives.

Selective media promotes the growth of specific bacterial strains, while it suppresses the growth of others due to the inclusion of particular substrate.

Table 1.1 Composition of Selective Enrichment Medium for Growth of Extremophiles

S. No.	*Component*	*Amount*	*Function of component*
1.	Casamino acids	7.5 g	Source of amino acids, N, S and P
2.	Yeast extract	10.0 g	Source of growth factors
3.	Trisodium citrate	3.0 g	C and energy source
4.	KCl	2.0 g	K^+ source
5.	$MgSO_4 .7H_2O$	20.0 g	S and Mg^{++} source
6.	$FeCl_2$	0.023 g	Fe^{++} source
7.	NaCl	250 g	Na^+ source for halophiles and inhibitory to non-halophiles
8.	Distilled water	1000 mL	

- Set pH 7.4

Minimal media contains the essential nutrient for the growth of bacterial strains which include inorganic salts, carbon sources, i.e., carbon-dioxide, water and glucose.

Table 1.2 Composition of Minimal Medium for the Growth of a Bacterium

S. No.	Component	Amount	Function of component
1.	Sucrose	10.0 g	C and energy source
2.	K_2HPO_4	2.5 g	pH buffer; P and K source
3.	KH_2PO_4	2.5 g	pH buffer; P and K source
4.	$(NH_4)2HPO_4$	1.0 g	pH buffer; N and P source
5.	$MgSO_4$ $7H_2O$	0.20 g	S and Mg^{++} source
6.	$FeSO_4$ $7H_2O$	0.01 g	Fe^{++} source
7.	$MnSO_4$ $7H_2O$	0.007 g	Mn^{++} Source
8.	Distilled water	1000 mL	

• Set pH @ 7.0

Differential media are those which allow identification and isolation of microorganism usually through the use of physiological reaction (externally visible) important to the bacteria. The most important feature of these types of media is their ability to select and differentiate common microorganisms, e.g., MacConkey agar is a selective as well as differential media. It is selective for Gram-negative bacteria and distinguishes such types of bacteria which have the capability to ferment lactose and other sugars. Eosin methylene blue agar (EMB) belongs to the same class for detection and isolation of Gram-negative intestinal pathogens.

Enrichment media allows identification of fastidious microorganisms because of addition of peculiar growth factors. Enrichment cultures are those which are obtained with the use of selected media and altering the incubation conditions to isolate the microorganisms from environmental samples. Further, other factors which include relative growth factors or inhibitory effects which may change the result of enrichment culture, however the phenomenon is collectively known as **enrichment bias**.

Table 1.3 Composition of a Defined Medium for the Growth of a Bacterium

S. No.	Component	Amount	Function of component
1.	NH_4Cl	0.52 g	N source
2.	KH_2PO_4	0.28 g	P and K source
3.	$MgSO_4$ $7H_2O$	0.25 g	S and Mg++ source
4.	$CaCl_2$ $2H_2O$	0.07 g	Ca++ source

5.	Elemental sulfur	1.56 g	Energy source
6.	CO_2	5%*	C source
7.	Distilled water	1000 mL	
8.	pH 3.0		

*Aerate medium from time to time with air having 5% CO_2.

Composition of Some Important Media for Growth of Bacteria

Nutrient agar

Nutrient media is commonly used for the cultivation of non-fastidious organisms, which can be enriched with blood or other nutritious fluids. Nutrient agar is very suitable for demonstration and teaching purposes where survival of desired microorganisms is of utmost importance because at optimum temperature other microorganisms can grow in the presence of more nutritious substrate. Additionally, it can be used for the cultivation and identification of microorganisms which do not have any specific nutritional requirement. However, addition of different types of fluids which are of biological origin, such as blood of buffalo or goat, egg yolk and serum, etc. seems to be relatively suitable candidates for the cultivation of fastidious organisms.

Composition

- Peptic digest of animal tissue - 5 g
- Sodium chloride - 5 g
- Beef extract - 1.5 g
- Yeast extract - 1.5 g
- Agar -15 g
- Distilled water, add to make volume 1 L; adjust pH to 7.4–7.6

Peptic digest of animal tissue, beef extract and yeast extract provide the necessary nitrogen compounds, carbon, vitamins and also some trace ingredients necessary for the growth of bacteria. Sodium chloride maintains the osmotic equilibrium of the medium.

MacConkey agar

MacConkey agar is a medium developed to isolate culture and grow (Gram-stain negative) bacteria which distinguish them from those capable of fermenting lactose. However, there are other variants of MacConkey agar depending on the requirement of microbiologist and/or researcher.

Composition

- Peptone - 17 g
- Proteose peptone - 3 g
- Lactose - 10 g
- Bile salts - 1.5 g
- Sodium chloride - 5 g
- Neutral red - 0.03 g
- Crystal violet - 0.001 g
- Agar - 13.5 g
- Distilled water, add to make volume 1L; adjust pH to 7.1–7.2

It contains **bile salts** (to inhibit most **Gram-positive** bacteria), **crystal** violet dye (which also inhibits certain Gram-positive bacteria), **neutral** red dye (which stains microbes fermenting lactose), **lactose** and **peptone**.

Marine agar

Marine bacteria are present in large numbers in nutrient sea water. They are very essential to the life cycle of all kinds of microbial diversity which is present in seawater. Further, the identification and activity of marine bacteria is very important for the food industry. The medium contains all the essential nutrients required for the growth of such kind of microorganisms. Media have minerals equivalent to those present in seawater, such as peptic digest of animal tissue and yeast extract as the sources of nutrients essential for the marine bacteria. High amount of salt content is used to imitate seawater. Additionally, other minerals are used to represent the mineral composition of seawater.

Composition

- Peptic digest of animal tissue - 5 g
- Yeast extract - 1 g
- Ferric citrate - 0.1 g
- Sodium chloride - 19.45 g
- Magnesium chloride - 8.8 g
- Sodium sulfate - 3.24 g
- Calcium chloride - 1.8 g
- Potassium chloride - 0.55 g
- Sodium bicarbonate - 0.16 g
- Potassium bromide - 0.08 g
- Strontium chloride - 0.034 g
- Boric acid - 0.022 g

- Sodium silicate - 0.004 g
- Sodium fluorate - 0.0024 g
- Ammonium nitrate - 0.0016 g
- Disodium phosphate - 0.008 g
- Agar -15 g
- Distilled water, add to make volume 1 L; adjust pH to 7.6–7.8

Trypticase soy agar

Trypticase Soy Agar (TSA) belongs to the category of multipurpose media which is used for culturing a vast range of microorganisms. They are used for a variety of purposes which include culture and subculture of microorganisms, storage of culture and isolation of pure cultures which are free of any contamination. TSA contains casein and soybean meal which act as a source of amino acids and other components which provide nitrogen to make it a wholesome media for a wide range of microorganisms. Glucose is the source of energy, while sodium chloride serves the function of buffer and maintains pH.

Composition

- Tryptone – 15 g
- Soytone (enzymatic digest of soybean meal) - 5 g
- Sodium chloride – 5 g
- Agar – 15 g
- Distilled water, add to make volume 1 L; adjust pH to 7.6–7.8

The media may be nurtured with additional ingredient, i.e., blood to promote the growth of bacteria having any specific requirement or any antibiotic source to allow the selection of different bacterial groups from the specified niche.

Brain heart infusion agar

Brain Heart Infusion Agar (BHIA) is an extremely nutritious medium which can support excellent growth of large number of microorganisms. However, it can further be enriched with additional ingredient, such as blood or rendered selective by different antibiotics. In conclusion, it is a general purpose media used for isolating a wide range of microorganisms from environment which are aerobic including clinical specimens.

Composition

- Calf brain, infusion - 200 g
- Beef heart, infusion - 250 g

- Proteose peptone - 10 g
- Dextrose - 2 g
- Sodium chloride - 5 g
- Disodium phosphate - 2.5 g
- Agar -15 g
- Distilled water, add to make volume 1 L; adjust pH to 7.4-7.6

Proteose peptone and infusions used in the media are the sources of carbon, while vitamins, amino acids and nitrogen serve as important growth factors. Dextrose is the source of energy. Sodium chloride maintains the osmotic equilibrium of the medium while, disodium phosphate acts as a source of buffer.

R2A agar

R2A agar is also known as Reasoner's 2A agar. The culture medium is commonly employed to study bacteria which are normally found in potable water. The use of such types of media which are low in their nutritional value can be helpful in isolating microorganisms which are either injured or under stressed conditions by providing longer incubation time. The bacteria which have a slow growth rate as compared to others which overcome their growth on a rich culture medium can be isolated using such types of media. Moreover R2A agar in comparison to other types of agar which includes tryptone glucose agar or standard method agar provides better recovery of microbial strains from stressed environment, i.e., hexachlorocyclohexane contaminated sites/hot springs/ acid mine drainage and chlorine tolerant bacteria from potable water systems.

Composition

- Casein enzymic hydrolysate - 0.25 g
- Peptic digest of animal tissue - 0.25 g
- Casein acid hydrolysate - 0.5 g
- Yeast extract - 0.5 g
- Glucose - 0.5 g
- Starch soluble - 0.5 g
- Dipotassium phosphate - 0.03 g
- Magnesium sulfate heptahydrate - 0.5 g
- Sodium pyruvate - 0.03 g
- Agar - 15 g
- Distilled water, add to make volume 1 L; adjust pH to 7.2-7.6

Enzymatic digest of casein, enzymatic digest of animal tissue, casein acid hydrolysate and yeast extract provide necessary nitrogen sources, carbohydrates, vitamins, minerals and growth factors to growing organisms. Dextrose serves as carbon source, soluble starch aids in recovery of injured organisms toxic metabolic by-products while, sodium pyruvate increases recovery of stressed cells. Magnesium sulfate is a source of divalent cations and sulfate. Dipotassium phosphate is used to balance the pH of medium.

Preparation of LB (Luria Bertani) broth (pH 7.0-7.5)

Learning objectives

To be able to

- Weigh chemicals using an electronic weighing balance
- Adjust the pH of solutions to required level using a pH meter
- Sterilize solutions and glassware/plasticware using an autoclave

Requirements

Glassware and Plasticware

- Beaker
- Measuring cylinder
- Culture flasks/test tubes (for LB broth)
- Schott Duran glass bottles (for LB agar)
- Microtips

Chemicals

- Tryptone
- Yeast extract
- Sodium chloride
- Glucose
- Agar powder
- Milli-Q water

Equipment

- Magnetic stirrer
- pH meter
- Weighing balance
- Micropipette
- Autoclave

Theoretical aspect/principle

Luria Bertani broth is an incorrect acronym for Lysogeny broth or Luria broth. It was first formulated by Giuseppe Bertani. It is a nutritionally rich medium especially in organic content provided by the tryptone and yeast extract in the media. The function of each of the components of the medium is as under:

Tryptone – It is a sourcc of amino acids, peptones and peptides required for the growth of bacteria.

Yeast Extract – It contains the vitamins and trace elements.

Glucose – Carbon source for the culture.

Sodium chloride – Sodium ions are necessary for transport and maintenance of osmotic balance.

Preparation

S. No.	*Components of LB*	*For 100 mL*	*For 250 mL*	*For 500 mL*	*For 1 L*
1.	Tryptone	1 g	2.5 g	5 g	10 g
2.	Yeast extract	0.5 g	1.25 g	2.5 g	5 g
3.	NaCl	0.5 g	1.25 g	2.5 g	5 g
4.	Glucose	0.1 g	0.25 g	0.5 g	1 g
5.	Agar powder	1.5 g	3.75 g	7.5 g	15 g

I. For preparing 1000 mL of LB Broth (liquid media):

1. Dissolve components of LB in ~800 mL of Milli-Q water (slightly less volume than finally required) in a beaker.
2. Adjust the pH to 7.0 to 7.5 with 1N NaOH using pH meter by constantly stirring the media.
3. Make up the volume of the media to 1000 mL using Milli-Q water.
4. Aliquot in desired quantities (according to the volume of culture required), in flasks/tubes plugged with cotton, and cover with foil.
5. Autoclave the LB broth flask or tubes prior to use.

II. For preparing LB Agar (solid medium to pour plates):

6. To the aliquot add agar (1.5%). For example, for 100 mL, 300 mL and 500 mL media add 1.5 g, 4.5 g and 7.5 g agar, respectively.
7. Autoclave the LB agar bottle prior to use.

Precautions

- Do not add agar prior to aliquoting because agar doesn't dissolve in water.
- LB broth and Agar plates/flasks/tubes should be stored at room temperature.

FREQUENTLY ASKED QUESTIONS

1. Why do we commonly use agar as a solidifying agent?
2. What are the basic precautions while preparing any growth medium?
3. Which medium is used exclusively for isolation of halo-tolerant bacteria?
4. Which medium is mostly used for standardization of physiological age of a bacterium for analysis of fatty acids?

REFERENCES

1. **Armisen, R., F. Galatas** (1987). Production, Properties and Uses of Agar, in *Production and Utilization of Products from Commercial Seaweeds* (Editor: Dennis J. McHugh) FAO corporate document repository.
2. **Bertani, G.** (2004). Lysogeny at Mid-Twentieth Century: P1, P2, and Other Experimental Systems. J. Bacteriol. **186**: 595-600.
3. **Perry, J., J. Stanley** (1997). *Microbiology Dynamics and Diversity.* Saunders College Publishing.

2

Growth and Maintenance of Culture

Vivek Negi
E-mail: *viveknegi2006@gmail.com*

INTRODUCTION

In molecular biology many microorganisms are involved in the experiment and we need specialized techniques to handle them. Growth media must have essential nutrients for proper growth and maintenance of culture. Autoclaving of the media and discard are directly and indirectly involved in good culture growth. While working with the microbes, issues related to biosafety and biohazard must be attended for personal safety and environmental health.

LEARNING OBJECTIVES

- Maintenance of microbial culture
- Autoclaving procedures
- Biohazard and biosafety

Maintenance of Microbial Culture

Agar Plates for Short-term Storage: Quadrant streak method must be used to streak on rich agar Petri plate in sterile conditions to get well isolated colonies. These plates can be stored in a refrigerator for about a month. Plates should be carefully examined for any contamination and if it is contaminated then it is discarded.

Spread plate/pour plate method: In spread plate methodology microbial culture is evenly distributing over LB agar plate.

- On the basis of the microbe select and prepare particular agar medium.
- To avoid condensation after pouring the plates, the autoclaved agar is cooled in the range of 45°C and 50°C. In a 100 × 15 mm media plate 15 to 20 mL of media is poured and the thickness of media is 0.3 cm approx.
- To avoid microbial contamination inoculum should be spread as soon as possible.
- Changing of pipette tips must be avoided while preparing serial dilutions to spread on the plate.

- A glass or metal spreader should be flame sterilized by alcohol (70% isopropyl or ethanol) after which it is allowed to cool.
- While spreading the inoculum on the surface of the plate a spreader is positioned in a particular angle and with even pressure, so that the inoculum is run along the length of the spreader.

Streaking and obtaining pure culture

As mentioned in the spread plate/pour plate method pure cultures are obtained by the given methodology. The pure culture of the microbe on the surface of a solid medium is as if a single cell occupies an isolated portion of the agar surface.

Subculturing microbial cultures

Subculture of microbes is the transfer of well grown culture to fresh newly prepared media. If the culture is grown from the stock (lyophilized or glycerol) then it is not subculturing.

Broth cultures for growth of large volumes

In volumetric flasks broth should be 1/5th of its volume to have oxygen in it with large surface: volume ratio. Master broth culture kept for overnight shaking must be dispensed in sterile test tubes under sterile conditions. It should be discarded after use.

Long-term maintenance of strain

- Although long-term storage of bacterial cultures is generally not recommended.
- While lyophilization is one of the best techniques for long-term storage.
- Another method to store the strain is at very low temperature (- 80°C) that includes cryoprotectant such as 20 per cent glycerol.

Instead of re-purchasing whenever needed, it would be convenient to maintain a stock of a pure culture. Most of them are relatively easy to maintain on an appropriate growth medium.

Requirements of stock culture

- use of an adequate amount of inoculum
- an appropriate culture medium
- an appropriate incubation temperature

Creating Microbial glycerol stocks for long-term storage

- After inoculating the microbes for overnight in liquid culture, add 500 μL of the overnight culture to 500 μL of 40 per cent glycerol in a 2 mL cryovial and gently mix.
- Store the glycerol stock tube at -80°C. The stock will be stable for years until subjected to subsequent freeze and thaw cycles.
- Scrape off some of the frozen bacteria of the top. Do not let the glycerol stock unthaw. Streak the microbes onto an LB agar plate for overnight.

Protocol for lyophilization process

- Prepare the sterile solution—compound, mix, filter
- Fill into vials
- Partially insert a rubber closure onto the vials
- Aseptically load the vials into a freeze dry chamber
- Freeze every single solution in every vial below a pre-determined critical temperature
- Use appropriate application of temperature and pressure to sublime ice from the product
- Further apply temperature and pressure to remove the necessary amount of bound water from the product
- Automatically stopper the vials, neutralize the chamber.

Tips for maintaining strains

- Streak the culture on agar plate as soon as you receive the plate.
- Refrigerate the plate only when the good visible growth of culture is seen.
- Plates and slants are preferred over broth culture.
- Always start seed culture from single bacterial colony.
- Subculturing is preferred once in a week for maintaining broth culture.
- If the strain looks contaminated, streak and maintain fresh culture.
- Restreak the cultures on slants/plates about once a month.

Autoclaving

An autoclave is used to sterilize materials to remove all forms of microbial life with the aid of high temperature and pressure. Effective use of autoclaving is must for proper sterilization of the materials and safe use of autoclave.

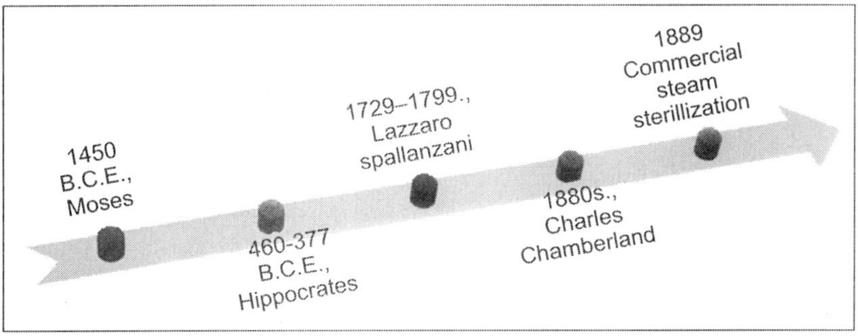

Fig. 2.1 Chronology of scientific advancement in autoclaving.

Historical significance (Fig. 2.1)

- The first sanitary code by Moses which included purification of materials by fire and boiling water.
- Hippocrates used boiling water to clean surgical instruments.
- Lazzaro Spallanzani discovered the protocol to kill bacteria in sealed glass flasks by heating it for 30 min.
- Pasteur had proposed boiling as the effective means of sterilization.
- Based on the Denis Papin invention of steam pressure cooker early design of the first autoclave was devised by Charles Chamberland.

Charles Chamberland

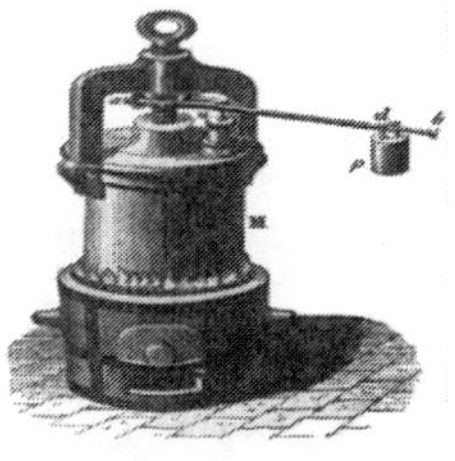

Papin steam cooker

Principle of autoclaving

- Steam penetrates materials in autoclave.
- Negative pressure is built due to condensation which draws in additional steam.
- Moist heat coagulates the proteins which kill the bacteria.

Two types of autoclave

- Gravity displacement.
- Vacuum/Gravity assisted.

Gravity displacement autoclave (Fig. 2.2)

- Cold air escapes through the bottom of the chamber as steam displaces it from above.
- The valves in this autoclave should never be obstructed and avoid overfilling of the chamber so that this system functions efficiently.

Vacuum/Gravity Assisted Autoclave:

- It pulls pressure from the chamber.

Manufacturers of autoclaves

- Steris
- Castle
- Getinge
- Tuttnauer

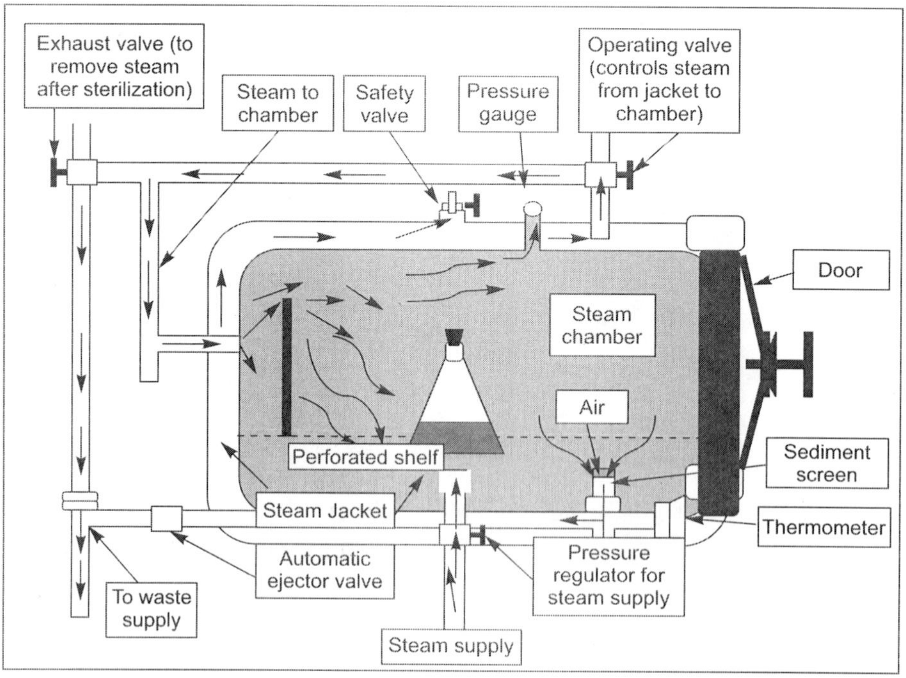

Fig. 2.2 Schematic diagram of the working of gravity displacement autoclave.

Standard pressure and temperature of an autoclave

When the free-flowing steam at a temperature of 100°C at pressure of 1 atmosphere above the sea-level pressure is subjected under pressure of 15 pounds pressure per square inch, the temperature inside the autoclave happens to rise up to 121°C, which is a common parameter employed in the moist heat sterilization.

Table 2.1 Some Standard Temperatures/Pressures Employed

S. No.	*Temperature (°C)*	*Pressure (psi)*
1	115	10
2	121	15
3	132	27

Psi = Pressure unit standing for pounds per square inch

Pre-Autoclave Procedures

(i) **Polypropylene bags**

They are autoclave or biohazard bags which are rigid, tear-resistant and available in various sizes. They should not be fully filled and some space should be left at the mouth region for passage of the air. Indicator tape should be placed in an "X" pattern over the biohazard bag.

(ii) **Materials that can be autoclaved**

- Plastic tubes, pipette tips and plates
- Glassware
- Surgical Instruments
- Biohazardous waste

(iii) **Risk Management**

Potential risks while operating autoclave

- Heat burns
- Steam burns
- Heat fluid scalds
- Hand and arm injuries due to door
- Body injury if there is any explosion

(iv) **Health and Safety**

- Trained person should be appointed as incharge of the autoclave.

- Autoclave should be operated only after wearing protective clothing.
- Autoclave must be inspected annually and proper record be maintained for its inspection, service and repairs.

(v) Personal Protective Equipment for Users

- Eye protection
- Heat-resistant gloves
- Lab coat, buttoned
- Closed-toed shoes

(vi) Loading Autoclave

- Place material in autoclave
- Do not overload
- The door should be firmly latched and closed

(vii) Steps to Autoclave

- Check that the water reservoir is filled.
- Set appropriate temperature for the cycle.
- Close and lock door.
- Pressure is achieved in 40 min approx. If autoclave is cold and 20 min approx. if in use.
- To STERILIZE turn on the autoclave.
- The autoclave should be turned to VENT when the cycle is complete, and door must remain closed until pressure drops. During venting temperature stays on for longer time and the steam present in chamber prevents melting of plastics.
- Turn off the autoclave.

Unloading Autoclave

- Heat-insulating glove provides protection against burns.
- Before opening the door one must ensure that the pressure of chamber is '8 psi'.
- While opening the door one should wear gloves and keep distance from the door.
- Super-heated liquid in containers should not be opened to avoid accidents.
- Autoclaved hot items should be placed in a room to cool the temperature.

Biohazard and Biosafety

For a microbiologist precautions must be taken in the laboratory to make sure that microbial infection or toxin that is harmful to human is not spread. Due to some major outbreaks, it is mandatory to have such facility at laboratory. International symbol for biological hazard (Fig. 2.3).

Fig. 2.3 International symbol for biological hazard.

Biohazard: An agent of biological origin that has the capacity to produce deleterious effects on humans, i.e., microorganisms, toxins and allergens

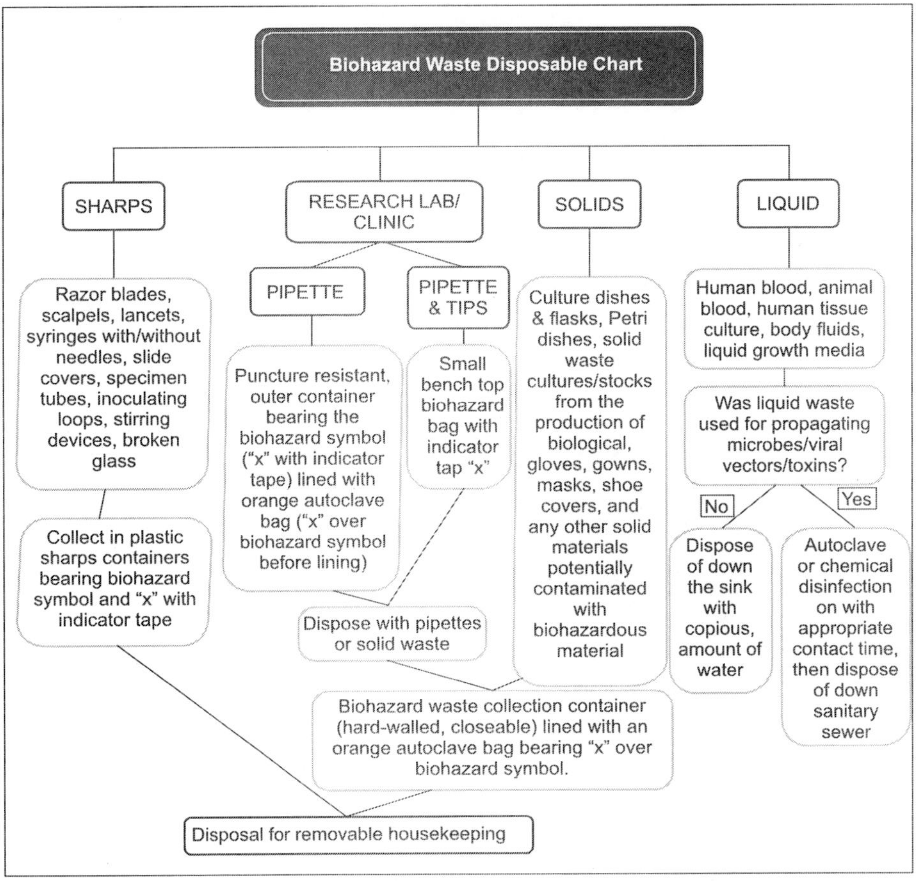

Fig. 2.4 Flow chart of the guidelines of solid, liquid and instrumental biohazard waste disposal.

derived from those organisms; and allergens and toxins derived from higher plants and animals.

Biosafety: The containment principles, technologies and practices that are implemented to prevent unintentional exposure to pathogens and toxins, or their accidental release.

FREQUENTLY ASKED QUESTIONS

1. What is the best way to preserve cultures for long?
2. Are there special requirements for reconstituting certain lyophilized microorganisms?
3. What should we do with contaminated plates?
4. How long should we sterilize the liquids?
5. Why contamination issues are observed even after autoclaving?

REFERENCES

1. **Thiel, T.** (1999). Introduction to Bacteria in Science in the Real World: Microbes in Action.
2. http://www.pasteur.fr/infosci/archives/chb0.html
3. http://www.astell.com/sterilisers
4. http://www.asu.edu/ehs/documents/autoclave-sop.pdf

3
Isolation of Genomic DNA

Udita Mukherjee
E-mail: *udita297@gmail.com*

INTRODUCTION

Genomic DNA is the chromosomal DNA of an organism and it encodes the genome of the organism. The isolation of genomic DNA is important as it serves as one of the important starting materials of many further experiments, such as PCR, restriction digestion and whole genome sequencing, etc.

LEARNING OBJECTIVES

To be able to

- Isolate genomic/chromosomal DNA from both Gram-positive and Gram-negative bacteria
- Preparation of crude lysates
- Purification of DNA and separation from co-isolating proteins and RNA
- Use benchtop cold centrifuge

REQUIREMENTS

Glassware and Plasticware

- Conical flasks
- Centrifuge tubes
- Balance
- Microcentrifuge tubes (chilled)
- Dispensors
- Microtips

Chemicals, Buffers and Solutions

- GTE Mix
- Proteinase K (10 mg/mL)
- Tris-equilibrated phenol chloroform
- 10% sodium dodecyl sulfate (SDS).

- 5 M sodium acetate
- Isopropanol
- 70% ethanol
- Lysozyme solution
- Kirby's mix
- Sterilized deionized water
- Luria broth or any other appropriate media (100 mL) for the growth of bacteria till late log or early stationary phase.

Equipment

- Centrifuge with rotor (precooled at 4°C)
- Ice bucket with ice
- Incubator-shaker (maintained at 37°C)
- Water bath

PREPARATION

1. **GTE Mix**

   Working concentrations of 50 mM Glucose, 50 mM Tris-HCl pH 8.0, 10 mM EDTA pH 8.0 and volume made up with sterilized deionized water.

2. **Proteinase K**

   Proteinase K 10 mg dissolved in 1 mL of sterilized deionized water.

3. **Tris- equilibrated phenol chloroform**

   Melted phenol 100 mL added to 100 mL chloroform and equilibrated twice with 200 mL 0.1 M Tris-HCl pH 8.0 on a stirrer. Stored in a dark bottle.

4. **Sodium Dodecyl Sulfate (SDS) 10%**

   Dissolve 10 g SDS in 100 mL sterilized deionized water.

5. **5M Sodium Acetate**

   Dissolve 41 g of sodium acetate in 100 mL sterilized deionized water.

6. **Ethanol 70%**

   Add 70 mL absolute ethanol in 30 mL sterilized deionized water.

7. **Lysozyme Solution**

   Dissolve 2 mg lysozyme in 1 mL of sterilized deionized water.

8. **2X Kirby's mix**
   Add 2 g sodium triisopropylnaphthalenesulphonate, 2 g sodium-4-amino-salicylate, 5 mL 2M Tris-HCl pH 8.0 and 6 mL phenol and mix.

THEORETICAL ASPECT/PRINCIPLE

The isolation of genomic DNA (chromosomal/total DNA) is an important embodiment of the switch from cell biology to molecular biology. The DNA was first isolated by Friedrich Miesher in 1869, which was then known as the "nuclein". The genomic DNA isolation requires the separation of genomic DNA from RNA, protein, lipid, etc. The basic requisite for the isolation of genomic DNA is breaking up the cell's membrane using mechanical or chemical methods. Traditionally, the mechanical methods include physical cell lysis, mechanical disruption, sonication physical, homogenization, freeze/thaw cycles, etc.

However, for most of the bacterial DNA isolation, chemical methods of cell lysis are preferred. The next things to be kept in mind are purity, yield and integrity of the genomic DNA. These factors become more important, because there are several polysaccharides, RNA and proteins that co-purify with the genomic DNA and contaminate it thus compromising with its purity. However, most of the further applications of genomic DNA require its pure form.

Although there are many protocols for the isolation of genomic DNA, but none of them provides a complete solution to all these problems. It is also imperative to mention here, that the source of DNA isolation is an important consideration for the protocol to be followed and the quality as well as the purity of the product obtained. Since this book deals with bacterial gene manipulation, therefore, only the bacterial sources would be considered here.

We know that bacteria can be classified as Gram-positive and Gram-negative, based upon the cell wall characteristics. Therefore, the methods of cell lysis would be different for these classes of bacteria. In layman terms, the methods of cell lysis would be harsher for Gram positives (pertaining to thicker cell walls) and milder for Gram negatives (pertaining to relatively thinner cell walls). As mentioned in the prelude, *Escherichia coli* will be taken as a standard example of Gram-negative bacteria, while *Amycolatopsis mediterranei* will be taken as an example for Gram-positive bacteria.

As discussed above, there are many different techniques that are available for isolating the genomic DNA. But none of them is singly competent enough to produce pure genomic DNA. Therefore, combinations of different techniques are used, depending upon the further experimental requirements. Broadly, separation of genomic DNA from the cell involves the following steps.

1. Disruption
2. Lysis
3. Removal of proteins and contaminants
4. Recovery of DNA

Role of chemicals

- **GTE Mix:** It is mainly used for resuspending the cell pellet prior to lysing them and extracting the DNA. Besides GTE also provides a stable environment for the DNA. 50 mM glucose is added because it maintains osmolarity and it being a non-electrolyte doesn't interfere with the buffering properties of the other components. Tris pH 8.0 is the pH buffer and it maintains the near physiological pH which doesn't harm the components inside the cell. EDTA chelates the metal ions such as zinc, magnesium and calcium and thus prevent the degradation of the DNA.'
- **Proteinase K:** Proteinase K is a protease that is used to digest cell surface proteins and other proteins that are present inside the cells, and are then co-isolated with the DNA. Therefore, proteinase K prevents initial protein contamination in genomic DNA isolation.
- **Tris-equilibrated phenol chloroform:** Genomic DNA isolation mainly relies on biphasic separation. Tris-equilibrated phenol chloroform is basically a buffer saturated phenol chloroform solution. TrisHCl pH 8.0 is the typically used buffering agent in this experiment. Buffer saturated phenol has higher density than water, whereas chloroform is much denser than both water and buffer saturated phenol. Phenol acts as the inorganic phase and chloroform acts as the organic phase. Chloroform prevents phase inversion. This mixture is used during separation of the genomic DNA and then after centrifugation, the DNA separates out in a separate phase with phenol (upper phase) while the protein contaminants are separated out with chloroform in the lower phase.
- **Sodium Dodecyl Sulfate 10%:** SDS is a strong anionic agent and it removes the negative ions from the proteins and destroys its structure. This also aids in destroying the proteins of the cell membrane and destroys them, as a result of which cell lysis takes

place. It thus exposes the chromosomal DNA. SDS also frees the DNA from the histones and other DNA binding proteins attached to the DNA by denaturing them.

- **Sodium Acetate 5 M:** It is used to precipitate solubilized high molecular weight DNA from the phenol (inorganic layer).
- **Ethanol 70%:** It is used to wash/purify the DNA once it is isolated. Ethanol 70% maintains the osmolarity of the DNA.
- **Lysozyme Solution:** Aids in the lysis of the cell membrane and releasing the genomic DNA.
- **Kirby's Mix:** A mixture of different components, all isotonic to the DNA, but aid in the lysis of the cell membrane, destroy cellular proteins and aid in releasing the genomic DNA.

PROCEDURE

The methods used for the isolation of genomic DNA are different for Gram-negative and Gram-positive strains of bacteria.

Isolation of Genomic DNA From Gram-Negative Bacteria

Source: *Escerichia coli*

1. Overnight culture 2 mL was taken and the cells were harvested by centrifugation for 10 min.
2. GTE 875 μL mix was added to the cell pellet and the cells were resuspended in the buffer by gentle mixing.
3. SDS 100 μL of 10% and 5 μL of proteinase K were added to the cells.
4. The above mixture was mixed well and incubated at 37°C for an hour in an incubator.
5. One mL of phenol-chloroform mixture was added to the contents, mixed well by inverting and incubated at room temperature for 5 min.
6. The contents are centrifuged at 10,000x g for 10 min at 4°C.
7. The highly viscous jelly-like supernatant was collected using cut tips and was transferred to a fresh tube.
8. The process was repeated once again with phenol-chloroform mixture and the supernatant was collected in a fresh tube.
9. Sodium 100 μL of 5 M acetate was added to the contents and was mixed gently.
10. Isopropanol 2 mL was added and mixed gently by inversion till white strands of DNA precipitate out.

11. The contents were centrifuged at 5,000x g for 10 min.
12. The supernatant was removed and 1 mL 70% ethanol was added.
13. The above contents were centrifuged at 5,000x g for 10 min.
14. After air drying for 5 min 200 μL of TE buffer or distilled water was added. The DNA was incubated overnight at 37°C or so to dissolve completely in a water bath. The remaining samples are stored for further experiments.

Isolation of Genomic DNA From Gram-Positive Bacteria

Source: *Amycolatopsis mediterranei*

Kirby's mix protocol for isolation of genomic DNA

This method of isolation was reported by Kieser *et al.* (2000) and depends on lysis with phenolic detergent mix and phenol/chloroform extraction.

1. Inoculated the culture in appropriate media (50 mL) using one colony of the bacteria. The culture was incubated at an appropriate temperature with rigorous shaking for optimal growth. It was harvested till it reached late log or early stationary phase.
2. The pellet was resuspended in 3 mL TE25S buffer and added lysozyme solution (2 mg/mL).
3. The mix was incubated at 37°C for 10 min.
4. Added 4 mL of 2X Kirby's mix and agitated for 1 min on a vortex mixer. Gentle vortexing was preferred so that the DNA is not sheared.
5. Phenol/chloroform/isoamyl alcohol 8 mL was added and it was vortexed again for 15 s.
6. Centrifuged at 3500x g for 10 min.
7. Transferred the upper aqueous phase to a fresh tube containing 3 mL phenol/chloroform/isoamyl alcohol and 600 μL of 3M unbuffered sodium acetate.
8. Vortexed the mix and centrifuged at 3500x g for 10 min.
9. Added 0.6 volume of isopropanol, invert mixed and observed thread-like structure formation. These DNA threads were spooled out and given a wash in ethanol 70%.
10. The pellet was kept for air drying and redissolved in sterilized deionized water.
11. The DNA was then quantified and was checked using agarose gel electrophoresis.

RESULTS

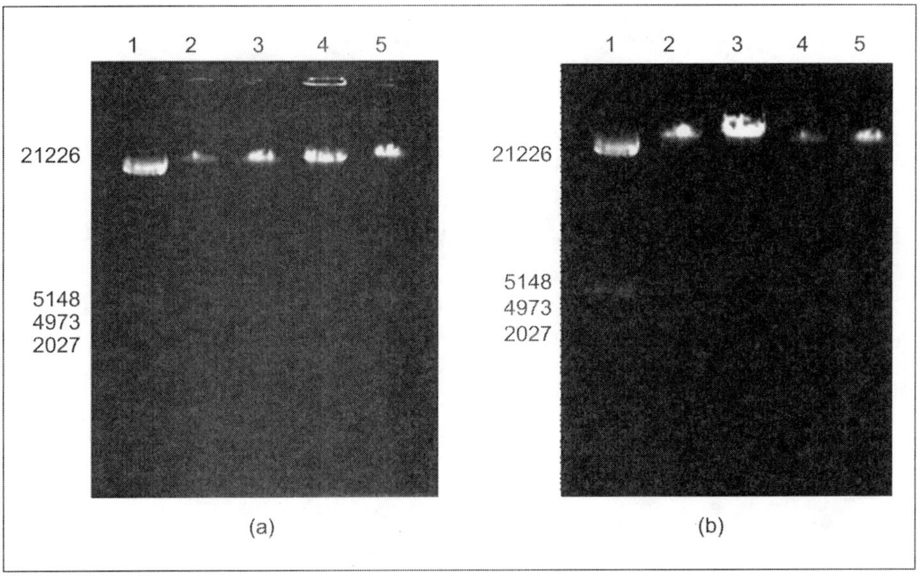

Fig. 3.1 0.8% agarose gel electrophoresis of genomic DNA isolated from (A) *E. coli* (B) *A. mediterranei* S699. The DNA was purified and RNase treated. (a) Lane 1: Lambda DNA/*Eco*RI/*Hind*III marker (Fermentas, SM0191). Lane 2, 3, 4 & 5: Genomic DNA isolated from *E. coli* (Gram –ve bacterium). (b) Lane 1: Lambda DNA/*Eco*RI/*Hind*III marker (Fermentas, SM0191). Lane 2, 3, 4 & 5: Genomic DNA isolated from *A. mediterranei* S699 (Gram +ve bacterium).

DISCUSSION

The genomic DNA was isolated, dissolved in sterilized deionized distilled water. It was then subjected to purification treatment with RNase, to remove RNA contamination. It was then electrophoresed on 0.8% agarose gel electrophoresis and was run overnight, at very low voltage (~30V) for better resolution of the DNA and bands of the Lambda *Eco*RI/*Hind*III ladder. The gel was then exposed to UV light in a gel documentation system, where the fluorescing DNA bands were seen, (Fig. 3.1) and the gel picture was saved. The DNA can be used for further experimentation.

FREQUENTLY ASKED QUESTIONS

1. What is the difference in the procedures between the isolation of genomic DNA from Gram-negative and Gram-positive strains of bacteria?

2. What are the basic considerations to be kept in mind while lysing the cells during the procedure? What effect would longer lysis incubation have on the genomic DNA?

REFERENCES

1. **Green, M.R., J. Sambrook** (2012). Isolation of High-Molecular-Weight DNA Using Organic Solvents to Purify DNA, in *Molecular Cloning: A Laboratory Manual*. 4th Ed. Cold Spring Harbor Laboratory Press, New York.
2. **Kieser, T., M.J. Bibb, M.J. Buttner, K.F. Chater, D.A. Hopwood** (2000). Practical Streptomyces Genetics. John Innes Foundation, Norwich.

4

Isolation of Plasmid DNA

Ankita Dua & Udita Mukherjee
E-mail: *ankitadua1987@gmail.com*
udita297@gmail.com

INTRODUCTION

The term 'plasmid' was coined by Joshua Lederberg in 1952, referring to the extrachromosomal DNA of an organism. It is the extrachromosomal DNA of an organism capable of self-replication as well as horizontal transfer, and confers certain characteristics to the bacterium, such as antibiotic resistance, etc.

LEARNING OBJECTIVES

To be able to

- Isolate plasmid DNA using three different methods:
  - (i) Alkaline lysis method
  - (ii) Kado and Lui's method
  - (iii) Hansen and Olsen's method (specific to megaplasmids)

THEORETICAL ASPECT/PRINCIPLE

Plasmids are small, circular, double-stranded DNA molecules distinct from a cell's chromosomal DNA. Plasmids naturally exist in bacterial cells, and they also occur in some eukaryotes as well.

Depending upon the strand of plasmids they can be:

(a) Covalently closed circular (CCC) – if both strands are intact circles
(b) Open circular (OC) – if only one strand is intact
(c) Linear (these are exceptions found in some bacteria; e.g., *Streptomyces* sp.)

Plasmids can be classified as conjugative or non-conjugative depending upon the presence or absence of "tra" genes that permit bacterial conjugation. Also copy numbers of plasmids vary and it also defines the ease with which it can be isolated. Relaxed plasmids have a higher copy number as compared to stringent plasmids which have a limited number.

Plasmids encode a few proteins for their own replication and the rest are encoded by the host cell. Replication proteins bind to the sequence called "ori" which is the origin of replication. The host range of the plasmid is determined by its "ori".

Genes present on the plasmid often hand over some special characteristics to the host cell such as antibiotic resistance. Plasmid sizes can vary from 1 kb to several hundreds of kb. Hence, depending upon their sizes they can be classified as miniplasmids and megaplasmids (>100 kb). Megaplasmids due to their sizes are more difficult to isolate and specialized methods are used to do so.

The total genomic DNA that is present in any cell is a total of the plasmid as well as chromosomal DNA. Hence, the plasmid needs to be separated from the large chunk of chromosomal mass. Differences in size and conformations of plasmids facilitate these methods of isolation of these extrachromosomal elements.

Three different methods of plasmid preparation are dealt with here.

(a) Alkaline lysis method

Also known as a plasmid mini-preparation method, this is widely used for making small preparations of plasmid DNA from culture volumes as low as 1ml.

Various chemicals are used in this protocol at distinct steps:

Mix I contains a mixture of glucose-Tris-EDTA and all the three components have indispensable functions in the protocol:

- Osmolarity of the DNA is maintained by addition of glucose. However, it has no role in maintenance of overall cell osmotic balance as these will be lysed eventually.
- Tris-HCl provides the alkaline pH 8.0 that protects acid labile DNA.
- EDTA chelates divalent metals, such as magnesium ions that help in preserving the overall cell wall structure and also inhibiting DNases at the same time (which may degrade DNA).

Mix II contains sodium hydroxide (NaOH) and sodium docdecyl sulfate (SDS, anionic detergent) which act together at pH as high as 8.0 provided by the former and solubilizes cell membranes and denatures chromosomal DNA and proteins, releasing plasmid DNA into supernatant.

Though this alkaline solution also denatures closed circular plasmid DNA, the strands of plasmid DNA do not separate from each other as they are topologically intertwined.

The **final solution** added is of potassium acetate (KOAc) and it acts to neutralize the pH which leads to renaturation of the two strands of plasmid DNA. The chromosomal lump however is too big to renature and hence forms a tangled mass.

Further, sodium ions are replaced by potassium ions (from potassium acetate) and hence the bacterial proteins, denatured chromosomal DNA get entangled and coated with SDS. The broken cell walls can be easily precipitated and removed by centrifugation. The insoluble salt of KDS is formed on interaction of KOAc + SDS. This facilitates removal of SDS from the plasmid DNA strands. The plasmid remains soluble in solution; however, larger genomic DNA is trapped by SDS which is further removed from solution by centrifugation.

Ethanol or Isopropanol: The excess KOAc salt still present in the supernatant will help the precipitation. As the DNA is negatively charged, in presence of alcohol the salt will mask the negative charges and allows the DNA to precipitate. This places the DNA in the pellet. Precipitated plasmid DNA is collected by centrifugation.

Precipitation with isopropanol is typically carried out using 0.6-0.7 volume of alcohol, while ethanol precipitation of DNA requires 2-3 volume of alcohol. Since less alcohol is required for isopropanol precipitation, it is given higher preference for DNA precipitation from higher volumes.

(b) Plasmid isolation using Kado & Liu's method

This procedure was specifically used to isolate plasmids of various sizes in species of *Agrobacterium*, *Escherichia*, and Rhizobium, etc. Covalently closed circular molecules are lysed from the cells using the combination of elevated temperature and detergents. Other cellular remnants would be discarded by carrying out extraction by phenol chloroform.

Under these conditions chromosomal DNA concentrations were reduced or eliminated. The extract obtained at this instant was used directly for electrophoresis. These procedures can also be used for selective isolation of plasmid DNA that can be used directly in molecular biology experiments, such as nick translation, restriction endonuclease treatment, transformation and DNA cloning.

(c) Plasmid isolation using Hansen and Olsen's method

This method was directly targeted as an improvement to all steps to prevent shearing of large megaplasmids and avoid their loss due to co-precipitation with chromosomal DNA. Strict lysis conditions were defined in the method that was followed to release the plasmid DNA.

- Cell lysis was performed using lyzosyme, disodium EDTA (Na_2EDTA) and SDS to a final concentration of 4% at a temperature of 55°C.

- Alkaline denaturation was performed at a pH as high as 12.6.
- Neutralization was done using Tris to lower the pH to 9.0.
- Membrane chromosome complexes were formed using a mixture of SDS and NaCl. These were removed as white flock material.
- Plasmid DNA was concentrated using PEG (polyethylene glycol) and finally resuspended in a very small volume to be used for visualization on gel.

A big highlight of the method is that all manipulations done were very gentle and precise, and involved no pipetting or harsh treatments. This completely avoided the loss of plasmids due to shearing. Also use of low temperature as well as high concentration of SDS would avoid any endonucleolytic activity on the plasmids and hence increase their yield as well.

REQUIREMENTS

Glassware and Plasticware

- Microfuge tubes
- Conical flasks
- Centrifuge tubes
- Balance
- Microcentrifuge tubes (chilled)
- Dispensers
- Microtips

Media

Luria broth or any other appropriate medium (100 mL) for the growth of bacteria till late log or early stationary phase. Also antibiotics are used for growth of plasmids under selective pressure (e.g., use of ampicillin for growth of pUC 18).

Chemicals, Buffers and Solutions

- Ampicillin (100 mg/mL)
- GTE mix (Glucose-Tris-EDTA)
- Sodium Dodecyl Sulfate (SDS 10%)
- 3 N NaOH
- 3 M potassium acetate (KoAc)
- Isopropanol
- Ethanol 70%

- Milli-Q water
- 1 M Tris pH 12.6
- Unbuffered phenol chloroform (1:1 v/v)
- Diethyl ether
- 5 M sodium acetate
- 1 M Tris-acetate
- 1 M sodium EDTA
- 1 M sodium phosphate buffer, pH 7.0
- 1 M sucrose
- Tris buffer pH 8.0
- Lysozyme (30 mg/ml)
- 2 M Tris pH 7.0
- 5 M NaCl
- Polyethylene glycol (PEG)
- Agarose powder
- 10X TBE
- Ethidium bromide
- Bromophenol blue

Special Equipment

- Centrifuge with rotor (pre-cooled at 4°C)
- Ice bucket with ice
- Incubator-shaker (maintained at 37°C).
- Water bath

PREPARATION

1. **Glucose-Tris-EDTA (GTE Mix):** Working concentrations of 50 mM Glucose, 50 mM Tris-HCl pH 8.0, 10 mM EDTA pH 8.0 and volume made up with Milli-Q water.
2. **Sodium Dodecyl Sulfate (SDS) 10%:** Dissolve 10 g SDS in 100 mL Milli-Q water.
3. **Sodium hydroxide (NaOH) 0.2 N:** From a stock concentration of 1 N NaOH, prepare 0.2 N NaOH.
4. **Neutralization Solution (3 M Potassium Acetate, CH_3COOK):** Dissolve 29.4 g of potassium acetate in 100 mL Milli-Q water.
5. **Ethanol 70%:** Add 70 mL absolute ethanol in 30 mL Milli-Q water.
6. **50 mM Tris pH 12.6:** From 1 M stock solution of Tris, prepare 50 mM Tris. Adjust pH using NaOH pellets.

7. **Unbuffered Phenol-Chloroform (1:1):** Mix equal volumes of phenol and chloroform.
8. **Sodium acetate 3 M:** Make using 5 M stock of sodium acetate.
9. **Tris-acetate 40 mM:** From 1 M stock Tris-acetate solution, prepare 40 mM solution and make up the volume with Milli-Q water.
10. **Sodium EDTA 2 mM:** From 1 M sodium EDTA solution, make 2 mM solution and make up the volume with Milli-Q water.
11. **10 mM sodium phosphate buffer pH 7.0:** For making 1 M sodium phosphate buffer (stock solution), make 1 M sodium phosphate (monobasic) and 1 M sodium phosphate (dibasic). Mix the two solutions and adjust the pH. Then from 1 M sodium phosphate buffer, make 10 mM solution and make up the volume with Milli-Q water.
12. **25% Sucrose in 50 mM Tris buffer pH 8.0:** Dissolve 25 g of sucrose in 100 mL 50 mM tris buffer pH 8.0.
13. **Lysozyme:** 10 mg/mL prepared using a stock of 30 mg/mL.
14. **Sodium EDTA 0.25 M** prepared from 1 M stock.
15. **42% PEG (w/v in 0.1 M sodium phosphate buffer):** 4.2 g PEG weighed and added to 10 mL of the buffer.
16. **TES buffer:** Made up of 0.5 M Tris pH 8, 5 mM EDTA pH 8.0 and 0.05 M NaCl.
17. **Preparation of agarose gel:** Refer to chapter 6 of the manual.

PROCEDURE

A. Alkaline Lysis Method to Recover Plasmid from *E. coli* Cells with Plasmid pUC18 (Fig. 4.1)

1. 5 mL of LB medium containing 100 μg/mL ampicillin was inoculated with an *E. coli* colony containing plasmid pUC18. It was incubated overnight at 37°C in orbital shaker.
2. 1.5 mL of overnight grown culture in a microfuge tube was taken and centrifuged at 5,000x g for 5 min at room temperature.
3. The culture was re-pelleted.
4. The supernatant was decanted, leaving the bacterial pellet as dry as possible. (It was gently tapped on a clean tissue paper.)
5. The pellet was resuspended in 200 μL of suspension solution (mix I/GTE mix) (ice cold) by vigorous tapping using knuckles or vortexer.

6. It was incubated at room temperature for 5 min.
7. 400 µL of alkaline lysis solution (Mix II: 1% SDS and 0.2 N NaOH) was added to the pellet and mixed by gentle inversion 5-6 times.
8. The mixture was incubated on ice for 5 min. (Time of incubation should not exceed, as prolonged exposure to denaturing conditions may cause plasmid DNA to enter an irreversibly denatured state.)
9. 200 µL of neutralization solution was added and mixed by inverting several times (solution will become viscous at this stage).
10. Incubated on ice for 10–15 min.
11. Centrifuged at 10,000x g for 20 min.

Recovery of plasmid DNA

1. The supernatant was transferred to a fresh tube and equal volume (650-700 µL) of isopropanol was added. It was mixed by inversion and incubated at room temperature for 5 min.
2. Centrifuged at 10,000x g for 20 min.
3. The supernatant was discarded by inverting the tube to allow all the liquid to drain out. It was decanted slowly in order to avoid pellet loss.
4. ~200 µL of ethanol 70% was added and re-centrifuged at 10000x g for 20 min.
5. Decanted the supernatant by inverting the tubes.
6. The pellet was air dried until no fluid is visible and the pellet turns transparent (overnight).
7. Dissolved in 50 µL of autoclaved Milli-Q water.
8. The plasmid DNA was visualized on 0.8% agarose gel.

B. Plasmid Isolation Using Kado & Liu's Method (Fig. 4.2)

1. Cells of the strain were grown overnight at 28°C at 200x in LB broth.
2. 2 mL culture suspensions were pelleted at 5700x g, 4°C for 7 min (done in triplicate for different lysis incubations).
3. Cells were lysed with the lysing solution (3% SDS & 50 mM Tris pH 12.6) and incubated at three different temperatures-55°C/65°C/95°C for 20 min (The temperature treatment is used to completely eliminate chromosomal DNA and can be extended as well.)

4. Solution was emulsified by shaking briefly on addition of 2 volumes of un-buffered phenol-chloroform (1:1 v/v) and centrifuged at 6000x g, 15 min, 4°C (clear aqueous phase contains the DNA; phenol solution lowers the pH of solution as well as separation of aqueous and hydrophobic phases, and remove proteins).
5. The aqueous DNA solution was freed of phenol by repeated extractions with diethyl ether and transferred to a new tube.
6. 100 mL of sodium acetate 3 M was added and the precipitation was done using 800 mL of ice cold ethanol and incubation at 20°C for 5 min.
7. Precipitated plasmid DNA was pelleted by centrifugation at 2°C for 15 min; air-dried and resuspended in 50 mL Milli-Q water.
8. Plasmids were checked on 0.7% agarose gel. Buffer system used in this protocol is a low salt system made up of 40 mM Tris–acetate and 2 mM sodium EDTA (to avoid overheating and allow resolution of high weight plasmids). Gel was resolved on a low voltage of about 20 V.

C. Plasmids Isolation Using Hansen & Olsen's Method (Fig. 4.3)

1. Cells of the strain were grown overnight at 28°C at 200x g in LB broth.
2. 50 mL culture was centrifuged at 5000x g for 15 min at 8°C.
3. The cell pellet was resuspended in 13 mL of sodium phosphate buffer (10 mM, pH 7.0) and centrifuged at 5000x g for 15 min at 8°C.
4. The cell pellet was resuspended in 2 mL of osmolarity solution (25% sucrose in 50 mM Tris buffer pH 8.0) and mixed properly. In this tube was added 140 mL of 10 mg/mL lyzozyme; mixed gently with 5 inversions and incubated on ice for 5 min.
5. 700 mL of sodium-EDTA was added (0.25 M pH 8.0) followed by 5 inversions and the tube was chilled for 5 min in ice.
6. 800 mL of SDS [20% w/v in Tris pH 8.0 and EDTA (10 mM: 1 mM)] followed by eight cycles of heat pulses and mixing (one cycle was 15 s in 55°C water bath followed by inversions for 15 s after removal from water bath) producing a clear viscous solution of lysed cells.

7. 700 mL of NaOH 3 N was added immediately, followed by 3 min of inversions.
8. 1.4 mL of Tris (2 M, pH 7.0) was added followed by inversion. The additions of 1 mL of SDS [20% w/v in Tris pH 8.0 and EDTA (10 mM: 1 mM)] immediately followed by addition of 2 mL of 5 M NaCl caused appearance of white flock materials. After inversion, tubes were chilled and incubated at 4°C overnight.
9. Tubes were centrifuged at 3000x g for 30 min at 4°C. The supernatant was transferred to a chilled tube and 0.313 vol of PEG 6000 (42% w/v in 0.1 M sodium phosphate buffer, pH 7.0) was added, mixed gently and incubated overnight at 4°C.
10. Tube was centrifuged at 3000x g for 5 min at 4°C that yielded pellet that was suspended in 200 mL of TES (Tris 0.05 M pH 8.0 and 0.05 M NaCl).

RESULTS

A. Using alkaline lysis

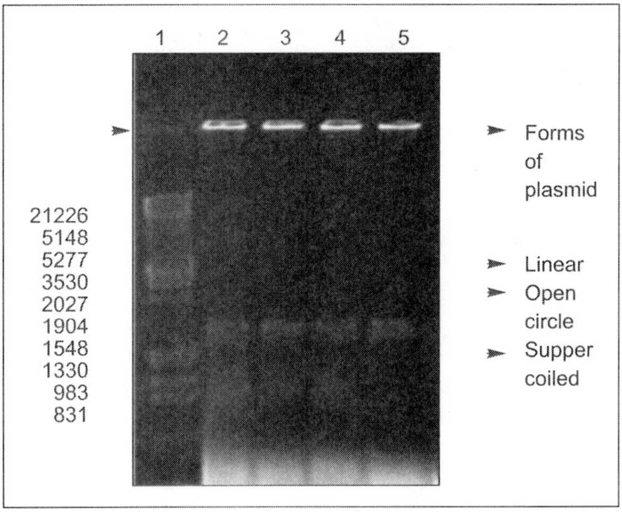

Fig. 4.1 0.8 % agarose gel electrophoresis profile of plasmid pUC18 from *E. coli*.
Lane 1: Lambda DNA/ EcoRI + HindIII marker (Fermentas, SM0191).
Lane 2, 3, 4 & 5: Plasmid pUC18 (2686 bp) in different conformations.

B. Using Kado & Lui's method

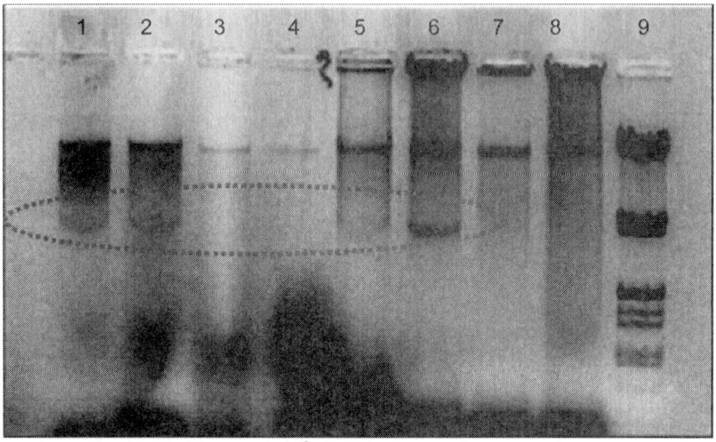

Fig. 4.2 0.8% agarose gel electrophoresis of plasmid DNA isolated from a bacterium. Lanes 1,2 and 5,6– plasmid DNA isolated at a temperature condition of 55°C and 95°C; Lanes 3,4 and 7,8– temperature conditions were varied to 45°C and 65°C. Lane 9: Gene ruler- Lambda DNA *Eco*RI/*Hind*III digested (21 kb to 500 bp).

C. Using Hansen & Olsen's method

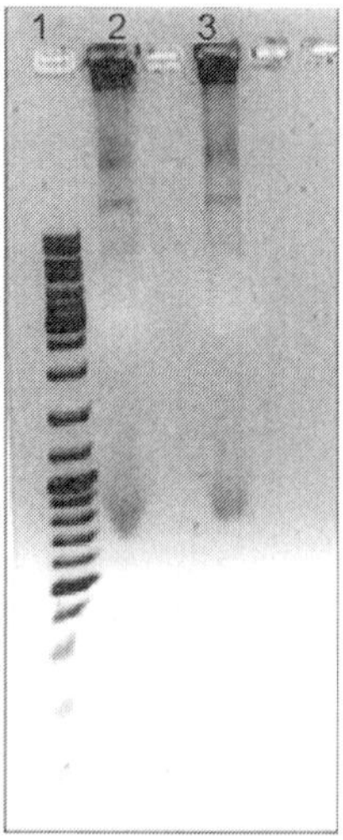

Fig. 4.3 0.8% agarose gel electrophoresis of plasmid DNA isolated from a bacterium: Lane 1: Gene ruler- 100 bp ladder; Lanes 2 and 3: Plasmid DNA isolated.

On following the methods of plasmid isolation, three conformations of the plasmid DNA are seen:

- The fastest moving is the supercoiled conformation, which is the native (in vivo) conformation and the last band seen on exposing the agarose gel under UV light.
- The next band above is the nicked circular band. This conformation has a cut on one of the strands of the plasmid DNA and it appears as a blobby mass on the agarose gel.
- The slowest and the topmost band to move is the linear band, which has cuts in both the strands of the plasmid DNA.

It must be mentioned that not all conformations of the plasmid are always visible on the agarose gel. It is in fact under rare circumstances, depending upon the size and nature of the plasmid, and of course on the method of plasmid purification that all three conformations are visible on the agarose gel. The bands that are most commonly seen are the nicked circular and the linear conformations. It is to be kept in mind that whenever a plasmid containing an insert is isolated and run on an agarose gel, a negative control (plasmid without the insert) must also be run along with, as it is serves as the initial indicator of whether cloning has been successful or not. The final result is seen through the difference in the mobility of the plasmids. The plasmid without the insert shall always move faster than the plasmid with the insert and majority of the times, it is clearly visible on the agarose gel.

PRECAUTIONS

1. Sterilized plastic and glassware should always be used.
2. Lysis should be checked before proceeding with the further renaturation. The mixture must turn slimy and should be clear visibly.
3. Over lysis should be prevented. As soon as the lysis occurs, renaturation solution should be added.

FREQUENTLY ASKED QUESTIONS

1. What is the basic difference in the protocols for the isolation of plasmid and genomic DNA?
2. Should different protocols be followed for the isolation of the plasmids depending upon the size?
3. What are the three conformations of a plasmid? Are all three conformations visible on an agarose gel?

REFERENCES

1. **Birnboim, H.C. and Doly, J**. (1979). A rapid alkaline extraction procedure for screening recombinant plasmid DNA. *Nucleic Acids Res.* **7:** 1513-1523.
2. **Primrose, S.B., Twyman R.M.** (2006). Principles of gene manipulation and genomics, 7th Edition. In; *Basic Biology of Plasmid and Phage Vectors*. Blackwell Publishing, UK.
3. **Kado, C.I. and S.T. Liu.** (1981). Rapid procedure for detection and isolation of large and small plasmids. *J. Bacteriol.* **145:** 1365.
4. **Hansen, J.B. and Olsen, R.H.** (1978). Isolation of large bacterial plasmids and characterization of the P2 incompatibility group plasmids pMG1 and pMG5. *J. Bacteriol.* **135:** 227-238.

5

Isolation of RNA

Vipin Gupta
E-mail: *guptavipin142@gmail.com*

INTRODUCTION

The extraction of RNA in molecular biology is the primary and frequently the most vital step in performing molecular experiments such as reverse transcription, real-time PCR, transcriptomic/metatranscriptomic analysis using next generation sequencing. Experimental setups like microarray analysis, digital PCR, northern analysis and complementary DNA library construction depend on the quality and quantity of extracted RNA. RNA gives the clear and true picture of expressed genes and functionality of the system, either a community or a cell. Cellular RNA constitutes three major ribonucleic acids, namely, rRNA, tRNA and mRNA. Among all three types of RNAs, mRNA is a link between the DNA and the functional protein synthesis and hence is a potential molecular tool to assess the cell functional activity. Nowadays, the trend is moving towards the cDNA conversion from RNA as cDNA is more stable and less prone to degradation. High quality RNA is needed for cDNA synthesis as well. Hence, isolation of high quality RNA is a basic necessity in molecular experiments in lab.

LEARNING OBJECTIVES

- To isolate total RNA from the bacterial samples
- To store and preserve isolated RNA

REQUIREMENTS

Chemicals

- DEPC-treated water
- TRIzol reagent
- Ice cold PBS
- Ethanol 70%
- Isopropyl alcohol
- Nuclease/RNase free water or 0.5% SDS
- Chloroform

PREPARATION

- Make sure that all the equipment and solutions are RNase free.
- DEPC treated solutions are highly recommended.
- All instruments along with work bench must be cleaned with RNAZap.

THEORETICAL ASPECT/PRINCIPLE

TRIzol is a all set chemical reagent, intended to isolate good quality total RNA along with DNA and protein from cells and various other samples like yeast, animal, plant, human or bacterial origin. Chemically, TRIzol is a monophasic phenol solution with guanidine isothiocynate and other proprietary constituents which facilitate the segregation of variety of RNA species of large and smaller molecular size. TRIzol allows instantaneous handing out of large number of samples and is an upgrading to single-step isolation of RNA method developed by Chomcynski and Sacchi.

RNA (ribonucleic acid) is a polymeric biological molecule present in cells. Structurally, it consists of a long single-stranded chain of phosphates and ribose units with the nitrogen bases adenine, guanine, cytosine, and uracil which are glycosidically bonded to ribose sugar.

Isolation of high quality RNA from bacterial cells/culture/suspensions or any other environmental source like bacterial mat/sediment or water sample is a crucial step to perform various molecular experiments in a lab. At the same time the task of RNA isolation is difficult because of its low stability and less concentration as compared to DNA per cell. Here we describe an easy and hassle-free crude total RNA isolation protocol from biological samples by using TRIzol method.

The basic principle of this method is based on theTRIzol reagent which maintains RNA integrity effectively as it is a potent RNase inhibitor, simultaneously dissolving and disrupting cell component during sample homogenization. After centrifugation, addition of chloroform splits the solution into two, aqueous and organic phases. RNA is exclusively limited to aqueous phase, while DNA is limited to organic phase only.

PROCEDURE

TRIzol RNA isolation Protocol Crude method

1. 50–100 mg of sample/cell suspension was mixed with the help of glass-teflonhomogenizer in 1–2 mL of TRIzol for 40–60 s.
2. Cell lysate was incubated for 2–3 min at room temperature (20–30°C).

3. 0.3 mL of chloroform was added for each 1 mL of TRIzol.
4. It was vortexed for 15 s and incubated for 5 min at room temperature.
5. Samples were centrifuged for 10 min at 5000x g.
6. Upper aqueous phase was carefully taken out from the tube by angling it at 45° from each sample into a fresh tube. Approximately, 500 µL of aqueous phase was recovered without any DNA contamination.
7. Equal amount of phenol:chloroform: isoamyl alcohol (25:24:1) was added to the recovered aqueous phase.
8. Vortexed thoroughly and spun for 10 min.
9. Aqueous phase was separated out and put into a fresh tube and then equal volume of chloroform was added.
10. Spun for 10 min at the 10,000x g.
11. Aqueous phase was transferred to a fresh tube.
12. 0.25 mL of isopropanol and 0.25 mL of 1.2 M Na-Citrate/0.8 M NaCl was added to each tube.
13. It was invert mixed gently and kept for at least 5 min on ice (4°C).
14. Spun for 10 min on 10,000x g.
15. The supernatant was discarded and pellet was washed with 70% ethanol.
16. The pellet was dried in hood for 5 min, complete dryness of pellet was avoided as it can decrease the yield of RNA.
17. RNA was dissolved in 30 µL of NF or RNA LATER solution. (Do not dissolve in more than 50 µL as it will decrease concentration.)
18. Immediately, RNA was stored in deep freeze (–80°C).

DNA STATEMENT

Isolation of RNA with TRIzol method generally gives contamination of DNA. To assess DNA contamination and its careful removal without RNA dilution, DNase treatment should be done to the isolated RNA before further proceedings.

1. 0.1 volume of DNase buffer was added to each RNA preparation.
2. It was gently vortexed.
3. 2 µL (4 units) of DNase was added to each RNA preparation.
4. It was gently vortexed and incubated for 25 min at 37°C.
5. DNase inactivation reagent mixed and 0.2 µL was added to each RNA preparation. (Do not increase DNase inactivation more than 4 µL in any case.)

6. Incubated for 5 min at RT with invert mixed at an interval of 1 min each.
7. Centrifuged for 5 min at 10,000x g and RNA was transferred to fresh tube for storage at deep freeze (–80°C).

Storage of Isolated RNA

Isolated RNA pellet, either for total or mRNA preparation, is resuspended in a solution. It is essential that RNA be poised and stored in a safe RNase-free atmosphere at – 80°C in one of the numerous RNA storage solutions premeditated specifically for this purpose.

1. The RNA storage solution (1 mM sodium citrate, pH 6.4)
2. 0.1 mM EDTA (in DEPC treated ultrapure water)
3. TE buffer (10 mM Tris-HCl, 1 mM EDTA, pH 7)

TE and 0.1 mM EDTA solutions are frequently specified in common RNA isolation and analysis protocol but RNA storage solution, a buffer, that delivers superior RNA steadiness than 0.1 mM EDTA or TE because it has two features that diminish base hydrolysis of RNA: low pH and sodium citrate which acts both as pH buffer and chelating agents.

RESULT AND DISCUSSION

As shown in Fig. 5.1, it is clear that even after DNase treatment removal of DNA contamination is difficult. Isolated RNA by this crude method may contain some impurities like DNA contamination, low yield, degraded RNA, etc. In that case, few suggested solutions can improve the quality and quantity of the isolated RNA. It is highly recommended not to dry RNA pellet completely as this will degrade RNA. Do not centrifuge RNA above 12,000x g as this will frail the RNA and be sure not to take any of the interphase (contains the DNA) with the aqueous phase. This may add color contamination to RNA in aqueous phase.

PRECAUTIONS

1. TRIzol should be used in fume hood only as it contains phenol (noxious and corrosive) and guanidine isothiocyanate (an irritant), and can be of health issue if not handled carefully.
2. Use of gloves, lab coat and safety glasses is necessary working with TRIzol, since its contact to eyes, skin or respiratory tract may cause severe burns to bare areas.
3. Use of RNAZap to clean all instruments and workplace to remove any RNases.

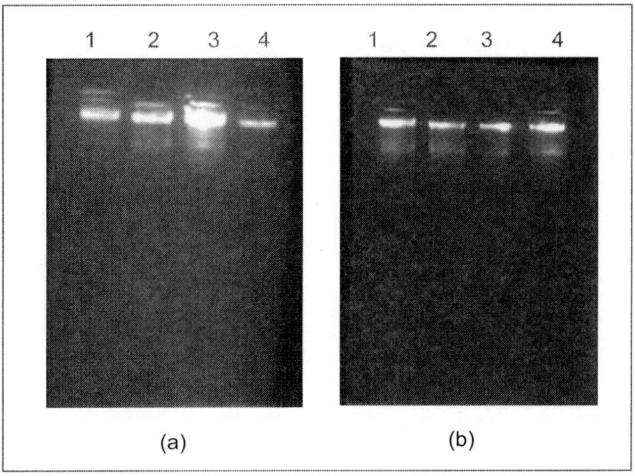

Fig. 5.1 (a) 2% agarose gel electrophoresis showing the isolated total RNA from community bacterial culture using the TRIzol method. DNA contamination is clearly visible in wells (1.2.3.4); (b) Samples with DNase treatment, showing reduced DNA contamination.

4. Don't dissolve RNA pellet in more than 50 μL.
5. Don't increase incubation time for DNase treatment, this will decrease RNA yield.

FREQUENTLY ASKED QUESTIONS

1. What amount of sample should be taken for good RNA output concentration?
2. What is the role of RNAZap in RNA isolation?
3. What are the drawbacks of using TRIzol reagent?
4. Why RNA isolation is to be done on ice (4°C) always?

REFERENCES

1. **Chomczynski, P., N. Sacchi** (1987). Single-step method of RNA isolation by acid guanidinium thiocyanate-phenol-chloroform extraction. *Anal. Biochem.* **1:** 156-159.
2. **Chomczynski, P., N. Sacchi** (2006). The single-step method of RNA isolation by acid guanidiniumthiocyanate-phenol-chloroform extraction: twenty-something years on. *Nat. Protoc.* **1:** 581-586.

6

Visualization of Nucleic Acids and Proteins

Ankita Dua, Helianthous Verma,
Princy Hira & Mallikarjun Shakarad
E-mail: *gankitadua1987@gmail.com*
helianthousverma@gmail.com
princy.hira@gmail.com
beelam.ms@gmail.com

INTRODUCTION

Nucleic acids and proteins post-isolation need to be resolved on gels to check for their quantity and quality. Electrophoresis is the method used in analytical biology for separation and visualization of proteins and nucleic acids. Protein gels will be discussed later in chapter 7 of this book.

LEARNING OBJECTIVES

To be able to

- Prepare agarose gels for resolving nucleic acids
- Run pulse field agarose gel equipment
- Load samples and resolve them

REQUIREMENTS

Chemicals

- Agarose powder
- Tris-borate EDTA buffer
- Milli-Q water
- LB broth/any culture medium for growth of bacterial culture
- Low melting agarose customized for PFGE
- Cell suspension buffer [Tris, NaCl and EDTA]
- Lysozyme buffer [Tris, NaCl, sodium deoxycholate, sodium lauryl sarcosine and lyzozyme]
- Proteinase K buffer [Tris, EDTA, sodium deoxycholate, sodium lauryl sarcosine and proteinase K]
- Wash buffer [Tris and EDTA]
- Buffer for restriction digestion

Equipment

- Power packs
- Gel tanks
- Pulse field gel equipment: electrode chamber, power module, coolant module and pump.

THEORETICAL ASPECT

The principle involved is to carry out separation of different sized molecules that carry a charge (like DNA is negatively charged) and migrate towards opposite charged electrodes. The separation depends on their varying sizes and mobility, hence attained when moving across the gel. The charge per unit length in any given fragment of DNA is identical and their mobility on moving towards the anode should also be similar under an applied electric field. Separation at this point is achieved due to the hindrance provided to the movement of molecules in the matrix of gel as a consequence of their varying sizes.

Larger molecules take longer to migrate across the gel pores as they experience more resistance, whereas the smaller ones remain unhindered. Double stranded DNA molecules exist as stiff rods and while passing through the pores, experience a drag and due to this, larger the molecule, more the retardation.

Gels can be of different types based on the length of molecules to be separated:

1. Horizontal agarose gels for separation of nucleic acids larger than 100 bp and up to ~10 Kb.
2. Polyacrylamide gels for short DNA segments (20-100 bp) and proteins.
3. Pulse field gel electrophoresis for separation of megabase sized DNA molecules such as megaplasmids and chromosomes.

1. Agarose gel electrophoresis of DNA

DNA molecules are separated on application of electric field, these are negatively charged molecules hence will migrate to the positively charged electrode "anode". Separation takes place through a matrix of a polysaccharide polymer agarose (extracted from seaweed) that is made up of repeating units of agarobiose (a disaccharide composed of D-galactose and 3,6-anhydro-L-galactopyranose). Gel concentration depends on the size of the molecule to be separated (Table 6.1).

Table 6.1 Size Selection of DNA that Can be Resolved at Various Concentrations of Agarose Gel

S. No.	*Gel Concentration*	*Size of DNA to be resolved*
1.	0.3%	dsDNA between 0.1 and 3 Kb
2.	0.8%	dsDNA between 0.5 to 10 Kb
3.	2%	dsDNA between 0.1 and 3 Kb

Conformation of the DNA molecules affects its migration, hence supercoiled DNA fragments move fastest due to their tight packaging. The most common concentration of agarose in gels used in daily procedures is 0.8%. DNA gels are run as horizontal, submarine or submerged gels and are completely immersed in the buffer. Buffers used can be of two types: Tris-acetate-EDTA (TAE) and Tris-borate-EDTA (TBE). Both buffers are commonly prepared as stock solutions of 50X and 10X respectively and further they can be diluted to make working solutions of 1X and 0.5X as per requirement.

TAE is less frequently used compared to TBE as it has lower ionic strength and buffering capacity. It needs to be replaced frequently or recirculated for longer gel run times (>4 h). TAE is used for high resolution of long nucleic acids (>1.5 Kb) and also nucleic acids move slower in these gels. Comparatively, TBE gels have a greater buffering capacity and also present a sharper resolution.

TAE Buffer

For the purpose of making 50X TAE buffer, the following amounts of chemicals are used:

242 g Tris base

57.1 mL glacial acetic acid

100 mL of 0.5 M EDTA (pH 8.0)

Volume is made up to 1 L. This solution can be further diluted to make working solutions of 0.5/1X.

TBE Buffer

For the purpose of making 10X TBE buffer the following amounts of chemicals are used:

108 g Tris base

55 g boric acid

40 mL of 0.5 M EDTA (pH 8.0)

Volume is made up to 1 L. This solution can be further diluted to make working solutions of 0.5/1X.

Staining of Gels

Staining of nucleic acid gels is done using dyes, such as ethidium bromide. It binds to the dye by insertion between the stacked base-pairs (intercalation) and fluorescence on illumination with ultraviolet light. EtBr runs in the direction opposite to that of DNA. Other methods of staining include using SYBR Green, methylene blue and crystal violet. SYBR Green is a safer alternative to EtBr as the latter has been reported to be carcinogenic and the former calls for blue-light transilluminator.

However, staining with crystal violet doesn't require the use of UV light, but the bands produced are not very intense.

Loading dye

Bromophenol blue is used as the loading dye and is mixed with the samples before loading on the gel. It is essential as it helps in tracking the movement of the sample and makes it heavier before loading. It migrates in same direction as that of a DNA fragment being approximately 500 bp ahead because of a negative charge at moderate pH. Its components are (for 100 ml 6X preparation):

70 g sucrose
10 mL EDTA (pH 8.0)
0.25 g bromophenol blue
Volume made up to 100 mL.

2. Polyacrylamide gel electrophoresis (PAGE)

Polyacrylamide gels are formed by polymerization of acrylamide monomer into long chains in a head to tail fashion. N,N′-bisacrylamide molecules (two acrylamide residues linked by a methylene group) act as a crosslinking agent as well as co-monomer in these gels and are responsible for chain extension by introducing secondary sites in growing chain. Ammonium persulfate and TEMED (N,N,N′,N′-tetra-methylenediamine) are required to initiate the polymerisation reaction. Chemistry of polymerisation is based on free radical catalysis, but photo-polymerisation using riboflavin can also be used as an alternative method. Temperature, concentration and purity of reagents should be carefully controlled to avoid non-uniform pore size as this process generates heat.

Like agarose gels, polyacrylamide gels can also have different pore sizes depending on the total concentration of acrylamide monomer and weight percentage of cross-linker present. PAGE is used for analysis of proteins (with the aid of SDS) and separation of DNA fragments in sequence analysis as these gels have smaller pore size and better resolution, i.e., it can separate DNA fragments differing in size by 1 bp, as compared to agarose gels. By varying concentration of acrylamide and bisacrylamide, an optimal pore size can be set to give better resolution and separation of nucleic acids as well as proteins. Separation in PAGE occurs due to relative mobility of charged species and frictional retardation based on size. More discussion about this technique has been dealt with in chapter- "Whole Cell Protein Isolation and SDS-PAGE" of this book.

3. *Pulse field Gel electrophoresis (PFGE)*

Scientists Schwartz and Cantor developed the technique of PFGE in the year 1983 as a method for overcoming limitations of resolving larger DNA molecules on standard agarose gels and allowing up to millions of base pairs in size to be fractionated.

In conventional DNA electrophoresis, molecular sieving is achieved in a size dependent manner. However, it does not work successfully for molecules longer than 50 Kb. Larger DNA molecules exist as ball-like random coils and to pass through smaller gel pores they need to unravel in a snake-like manner through the gel matrix.

During PFGE, electric field direction is altered periodically and hence, the DNA molecules acquire new orientations to pass through the gel matrix. Maximum electrophoretic mobility is achieved when the orientation of the molecules is with the direction of the given electric field.

The major components of a pulse field system are:

- **Power module:** For generation of electrode voltages and storage of function parameters.
- **Cooling module:** For maintenance of temperature (14°C).
- **Electrophoresis chamber:** Elimination of DNA lane distortion. (It is composed of 24 horizontal electrodes which allow reorientation angles of 60° or 120° due to their arrangement in a hexagonal fashion, in contrast to, two perpendicular electrodes in traditional gel systems.)

A number of configurations using currents applied in various directions have been developed.

- **Field inversion gel electrophoresis (FIGE)**

FIGE offers resolution of up to 800 Kb (600-750 Kb) fragments. Fields are 180° apart and forward sample migration is generated by reversal of polarity of electrodes at intervals with longer forward pulse time than reverse pulse time (Fig. 6.1a). Net forward migration is accomplished by increasing the ratio of forward to reverse pulse times to 3:1. The duration of pulse time is amplified gradually during a run to improve the resolution of the bands by FIGE. This phenomenon is known as "switch time ramping".

Alteration in pulse durations recurrently during the progress of an experiment in FIGE attributes to its advantages of straight lanes and simple equipment. It can be performed using conventional horizontal gel equipment supplemented with a pulse controller to change the field polarity.

- **Transverse alternating field electrophoresis (TAFE)**

This technique can resolve fragments up to 1,600 Kb fragments. The orientation of gel is vertical and a simple four-electrode array is placed in front and back, but not in the plane of the gel. Thickness of gel forces the sample molecules to zigzag, but the bands remain straight as when the molecules pass down the gel, all the lanes experience the same effect of continual variations in reorientation angles and field strength (Fig.6.1b).

However, the molecules do not move at a constant velocity as the angle between the electric fields varies over the length of the gel, i.e., 115° from the top of the gel to the bottom (approximately 165°).

- **Contour-clamped homogeneous electric field electrophoresis (CHEF)**

CHEF system is most frequently used in molecular techniques and provides resolution of 2-7 Mb DNA molecules. It is a setup of a number of electrodes that are outlining the gel (as in the positioning of a contour), which are clamped to specific voltages to produce a steady electric field. In this system, all the 24-point electrodes are equally spaced and arranged in hexagonal form, connected to power supply by means of resistors having same resistance and forming an external loop. This loop is responsible for setting the voltages of all the electrodes around the hexagonal contour to values appropriate to the generation of uniform fields in each of the alternate switching position (Fig. 6.1c).

The voltage is set at these 24 electrode points which produce reasonably uniform electric fields to accomplish all lanes of a gel to run straight. A reorientation angle of 120° is used in CHEF.

Best commercial PFGE systems have been developed by the company BIORAD by the name of CHEF DRI/II/III and Mapper.

Factors that affect resolution in PFGE

Electrophoretic parameters that affect resolution in PFGE are: pulse time, voltage, field strength, temperature, buffer composition, type and concentration of agarose and angle between the fields. Unlike horizontal agarose gels, pulse field gels are run at lower temperatures of 4-15°C. DNA migrates faster at higher temperatures; however, control of temperatures helps in controlling reproducibility of separation and

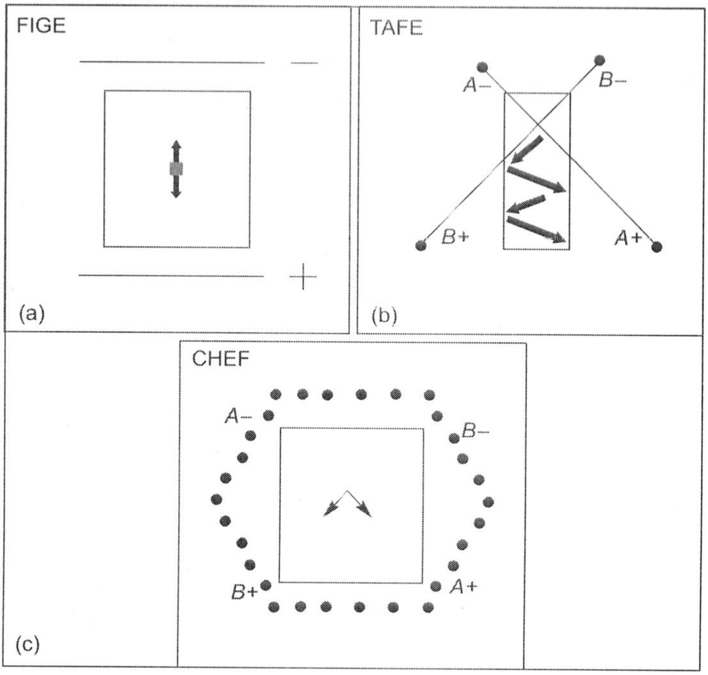

Fig. 6.1 Electrode configuration of commonly used pulsed field gel electrophoresis units: (a) FIGE, (b) TAFE and (c) CHEF
Source: Pulsed-field gel electrophoresis (Herschleb *et al.*, 2007)

mobilities. When running a particular sample of DNA, all these conditions need to be optimized to get the best results.

The DNA embedded in the plugs can be manipulated further by subjecting them to digestion with S1 nucleases or simple restriction enzymes. S1 nuclease is an apt enzyme to convert closed circular plasmids to linear forms that can be seen on gel and their sizes can be estimated.

Larger supercoiled plasmids migrate slowly compared to their linear counterparts and hence treatment with S1 nuclease is apt for size approximation. Along with the standard markers used in PFGE (ladders of *S. cerevisae*, *S. pombe* and λ DNA), the linear megaplasmids can be sized.

PROCEDURE

Casting of horizontal agarose gels

1. Preparation of 0.8% agarose gel: 0.8 g of agarose powder was dissolved in 100 mL of the desired buffer (most commonly 0.5X TBE).

2. The above solution was heated in microwave, cooled down to a temperature of 35°C and poured on to the glass/plastic gel casting tray. (The gel casting tray must be clean and also have a comb fit before pouring the solution.)
3. The gel was allowed to solidify for an hour before immersing into the buffer filled gel tank (The thickness and size of the cast gel depends on the amount of DNA sample to be loaded).
4. Appropriate amount of staining dye was mixed before pouring of the gel. (The stock concentration of EtBr = 10 mg/mL and the working concentration = 0.5 µg/mL were prepared.)
5. Once the gel was set, it was carefully transferred to the buffer filled tank making sure that it must be completely immersed in the buffer.
6. For loading, the sample was mixed with appropriate amount of loading dye, i.e., 6X bromophenol blue and loaded in the wells.

Performing pulse field gel electrophoresis

Pulse field gel electrophoresis is a lengthier technique in comparison to standard agarose gels as sample preparation, as well as run time takes about 2–3 days in total. The procedure is divided into three steps:

1. Growth of bacterial culture to appropriate cell density ($O.D._{600}$ of 0.6-0.8).
2. Preparation of cell culture in agarose plugs and cell lysis.
3. Loading the plugs onto the gel and running pulse field gel electrophoresis.

(a) Preparation of agarose embedded bacterial DNA

1. Bacterial culture was grown to appropriate $O.D._{600}$ of 0.6–0.8 in LB/appropriate culture media for preparation of culture plugs. (Different volumes of culture pellets can be made to test for best resolution on gel, i.e., 1 mL, 2 mL, 5 mL, etc. This needs to be standardized for different genera.) The cells were pelleted in microcentrifuge tubes (1.5 mL/2 mL) at 6000x *g* for 3 min.
2. Cells were resuspended in 500 mL of cell suspension buffer (10 mM Tris pH 7.2, 20 mM NaCl, 50 mM EDTA) and equilibrated at 50°C in water bath.
3. Two per cent low melting agarose (specially procured for running PFGE) was prepared using Milli-Q water, melted using a microwave and equilibrated at 50°C in water bath.
4. Cell suspension 500 mL was mixed gently with 500 mL of 2% low melting agarose (to make final concentration of 1% agarose). This

mixture was transferred to the plug moulds (80 mL volume) using sterile pipettes and allowed to solidify for 30 min at room temperature. It was later kept at 4°C for 30 min to solidify.

5. Post-solidifying, the plugs were carefully excised from the moulds and added to 1 mL of lysozyme buffer (10 mM Tris pH 7.2, 50 mM NaCl, 0.2% sodium deoxycholate, 0.5% sodium lauryl sarcosine and 1 mg/mL lysozyme) in a microcentrifuge tube and kept at 37°C for one hour for lysis without agitation.
6. The lysozyme buffer was removed from the microcentrifuge solutions and the plugs were washed with wash buffer (20 mM Tris pH 8.0 and 50 mM EDTA) along with simultaneous agitation (can be done using a gel rocker).
7. The plugs were then incubated with 1 mL of proteinase K buffer (100 mM EDTA pH 8.0, 0.2% sodium deoxycholate, 1% sodium lauryl sarcosine and 1 mg/mL proteinase K) at 50°C overnight.
8. The proteinase K buffer was removed from the microcentrifuge solutions and the plugs were washed with wash buffer four times along with simultaneous agitation. Plugs were stored at 4°C for further use.

(b) Digestion of the DNA embedded in agarose plugs

1. S1 nuclease digestion
    - Plugs were cut into equal 2–3 slices and were digested at 37°C for 10 min with 1U of S1 nuclease in 200 mL of S1 buffer [50 mM NaCl, 30 mM sodium acetate (pH 4.5) and 5 mM zinc sulfate]. Reaction was stopped by incubating the plugs in 100 mL of ES [1% N-laurylsarcosine, 0.5 M EDTA and 1 mg/ml proteinase K] buffer for 15 min.
    - The gel was run at appropriate parameters.
2. Restriction digestion of plugs using rare enzyme cutters
    - A master mix containing desired restriction enzyme concentration, its buffer and Milli-Q water need to be prepared. For PFGE, rare-cutting restriction enzymes should be used to get the minimum number of bands (15–20).
    - Then the plugs were incubated for 2–4 h at the appropriate temperature for the restriction enzyme to work properly. Then, to remove enzyme master mix, the digested slices were incubated with 200 mL 0.5X TBE buffer to inactivate enzyme.
    - The gel was run at appropriate parameters.

(c) Running Pulse Field Gel Electrophoresis

- Plugs were sliced to 1/3 their sizes, loaded onto 1% agarose gel (PFGE grade) and were further sealed using melted agarose and allowed to solidify.
- Meanwhile, fresh 0.5X buffer was prepared (~3 L) and poured into the PFGE gel chamber and pump was switched on.
- This allowed free circulation of buffer between the gel chamber and the cooling apparatus (gels are generally run at 14°C).
- Once the desired temperature was reached, the pump was turned off and the gel was placed onto the chamber.
- The apparatus was set on the optimum conditions (initial switch time, final switch time, volts, time and inclined angle).
- The pump was turned on followed by turning the run switch "on".
- At gel running conditions with consistent run time of 26 h, angle 106° and voltage gradient of 6 V/cm best profiles were seen for a number of genera, such as Acinetobacter, Sphingomonas, etc., however, the only variations were in pulse times: 25 s–45 s and 30 s–60 s. A voltage of 6 V/cm applies to a total voltage of ~ 200 V (6 X 33 cm, 33 cm is the migration distance). The gel running conditions need to be optimized for various genera.

(d) Staining the gel

- After the run was completed, the gel was transferred to staining tray containing 0.5X TBE buffer and EtBr (>0.5 µg/mL) was added uniformly into the buffer and the gel was kept for staining (with slight agitation on gel rocker).
- The gel was visualized under UV light after an hour and repeatedly post every half an hour interval to get better results.
- Destaining was done by immersing the gel in clean MQ water and then visualization was done as well.

RESULT

Post-running a horizontal agarose gel, we expect to see the DNA bands resolved according to their size and these are interpreted using the sizes of the standard marker used. A typical gel run to resolve amplified products of a 1500 bp gene is shown in Fig. 6.2. A very bright 1500 bp band is seen in all the samples and is calculated on matching with the marker.

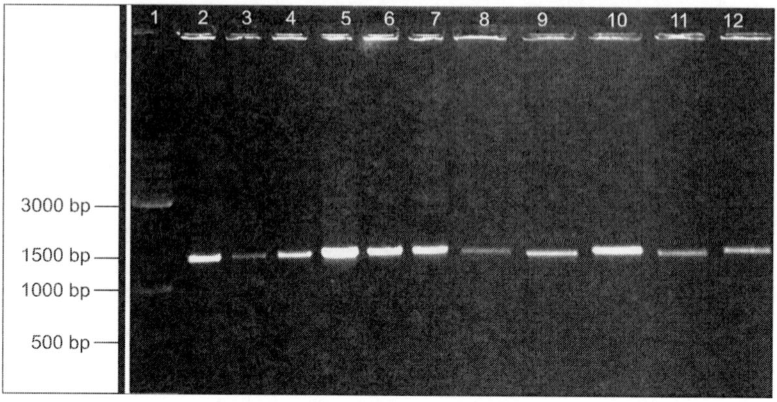

Fig. 6.2 0.8% agarose gel electrophoresis showing PCR amplified product of size 1500 bp. Lane 1: 100 bp ladder (Fermentas); Lane 2-12: amplified products.

DISCUSSION

Pulse field gels are usually run to resolve megabase sized DNA samples that cannot be resolved on a regular gel. Run time usually varies from species to species as separate pulses are applied in different dimensions.

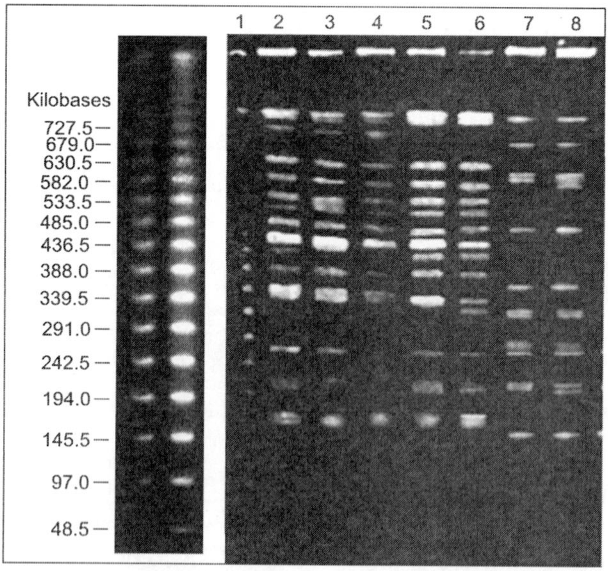

Fig. 6.3 Pulse field gel electrophoresis analysis of a 9-10 Mb strains of Amycolatopsis mediterranei: DNA restriction profiles *A. mediterranei* strains digested by AseI (pulse time: 5-90 s, run time: 36h, voltage: 200 V). Lane 1: λ DNA ladder PFGE marker; Lane 2-8: Strain DSM 43304, T-195, F1/24, Strain DSM 40773, Strain MTCC 17, Strain DSM 46096, Strain DSM 46095.

The expected resolution of digested profile of 9-10 Mb genomes of different strains of *Amycolatopsis mediterranei* can be seen in Fig. 6.3.

FREQUENTLY ASKED QUESTIONS

1. What is the least concentration of DNA that can be visualized using agarose gel?
2. What is the basic difference between horizontal agarose and pulse field gels?
3. Why is the enzyme S1 nuclease used?

REFERENCES

1. **Burmeister, M. & Ulanovsky, L. (Eds.)** (1992). *Pulse Field Gel Electrophoresis: Protocols, Methods & Theories*. The Humana Press Inc.
2. **Herschleb, J., G. Ananiev, D.C. Schwartz** (2007). Pulsed-field gel electrophoresis. *Nat. Protoc.* **2:** 677–684.
3. **Johnson, P.H., L.I. Grossman** (1977). Electrophoresis of DNA in agarose gels. Optimizing separations of conformational isomers of double and single-stranded DNAs. *Biochemistry* **16:** 4217–4225.
4. **Sharp, P.A., B. Sugden, J. Sambrook** (1973). Detection of two restriction endonuclease activities in Haemophilus parainfluenzae using analytical agarose-ethidium bromide electrophoresis. *Biochemistry* **12:** 3055–3063.
5. **Waring, M.J.** (1965). Complex formation between ethidium bromide and nucleic acids. *J. Mol. Biol.* **13:** 269–282.
6. **Wilson, K., J. Walker (Eds.)** (2010). Electrophoretic techniques, in *Principles and Techniques of Biochemistry and Molecular Biology*. 7th Edn., Cambridge University Press. 399–432.

7

Whole Cell Protein Isolation and SDS Page

Charu Tripathi & Namita Nayyar

E-mail: *charutripathi89@gmail.com*

nayyarnamita@gmail.com

INTRODUCTION

Isolation of whole cell proteins from bacterial cells is a basic technique that should be carried out before further purification and other downstream applications of purified proteins. The first step in protein extraction is lysis of the bacterial cell. Generally, cell lysis is brought about by disruption of the cell membrane by physical means (sonication) and in some cases by chemical means (detergents/enzymes). A centrifugation step then removes cell debris and the cell lysate or the protein containing part can be analysed by Sodium Dodecyl Sulfate Polyacrylamide Gel Electrophoresis (SDS-PAGE) under denaturing conditions provided by SDS.

LEARNING OBJECTIVES

To be able to

- Use a sonicator to bring about lysis of cells
- Understand the basics of protein isolation
- Understand the concept of SDS-PAGE
- Run SDS-PAGE under denaturing conditions

REQUIREMENTS

Glassware and Plasticware

- Conical flasks
- Centrifuge tubes
- Weighing balance
- Microcentrifuge tubes
- Dispensers
- Tips
- SDS-PAGE apparatus
- Power pack

Chemicals, Buffers and Solutions

- 30% acrylamide-bisacrylamide mix
- 1.5 M Tris-HCl (pH 8.8)
- 1.5 M Tris-HCl (pH 6.8)
- SDS
- 10% ammonium persulfate
- TEMED
- Methanol
- Acetic acid
- CBBR
- Tris base
- Glycine
- Imidazole
- NaCl
- β-Mercaptoethanol
- Bromophenol blue
- Glycerol
- 1% Agar
- Milli-Q water

Biological Material

- Bacterial culture *E. coli* DH5α

PREPARATION OF CHEMICALS AND BUFFERS

(a) 10X Laemmli (Tank Buffer): For 1 L

Table 7.1 Preparation of 10X Laemmli (Tank Buffer)

S. No	*Components*	*Amount*
1.	Tris base	30.3 g
2.	Glycine	144.1 g
3.	SDS	10.0 g
4.	Milli-Q water	1000 mL

- Do not autoclave
- Store at room temperature (RT)
- Working: 1X

(b) 30% Acrylamide-Bisacrylamide: For 200 mL

Table 7.2 Preparation of 30% Acrylamide-Bisacrylamide

S. No.	Components	Amount
1.	Acrylamide	58.4 g
2.	Bis-Acrylamide	1.6 g

- Dissolve components in 100 mL Milli-Q water initially and then make up the volume to 200 mL.
- Do not autoclave.
- Store at RT in a glass bottle covered in foil or in a brown bottle as it is light sensitive.

(c) 10% Ammonium Persulfate (APS): For 500 μL

- Add 50 mg APS to 500 μL Milli-Q.
- Do not autoclave.
- Store at 4°C. Use within 7 days.

(d) 10% SDS: For 100 mL

- Dissolve 10 g SDS in 100 mL Milli-Q water.
- Do not autoclave.
- Store at room temperature.

(e) 2X SDS GEL Loading Buffer:

Composition chart:

Table 7.3 Preparation of 2X SDS Gel Loading Buffer

S. No	Component	Concentration
1.	Tris-HCl pH 6.8	10 mM
2.	β-Mercaptoethanol	200 mM
3.	SDS	4%
4.	Bromophenol blue	0.2%
5.	Glycerol	20%

(f) 1% AGAR:

- Add 0.1 g agar to 10 mL autoclaved Milli-Q water.
- Heat in a microwave.

(g) CBBR Staining Solution:

- For 1 L: (To be prepared in fume hood)

- Dissolve 2 g CBBR in 500 mL methanol.
- Add 100 mL glacial acetic acid.
- Add 400 mL water.
- Do not autoclave. Store at room temperature.

(h) Destaining Solution: For 1 L (to be prepared in fume hood)

Table 7.4 Preparation of Destaining Solution

S. No.	*Components*	*Volume to be added*
1.	Methanol	400 mL
2.	Glacial acetic acid	100 mL
3.	Water	500 mL

Do not autoclave. Store at room temperature.

THEORETICAL ASPECT/PRINCIPLE

Electrophoresis is an important technique in molecular biology relying on the migration of charged molecules in solution when subjected to an electric field. Polyacrylamide Gel Electrophoresis (PAGE) separates proteins in the gel on the basis of shape, size and charge (Fig 7.1). Any charged molecule will migrate in a polyacrylamide gel under the influence of electric current, but the rate of migration will actually depend on its surface charge density. Thus, migration is dependent on not only the net charge, but also on the mass of the protein.

Acrylamide is a monomer. In the presence of free radicals, supplied by APS and stabilized by TEMED, a chain reaction is initiated in which monomers of acrylamide are polymerized into long chains. When the bifunctional agent N,N′-methylenebisacrylamide is included in the polymerization reaction, the chain becomes cross-linked to form a gel, whose porosity is determined by the length of the chains and the degree of cross-linking. If the pore size is large, then proteins of larger size can pass through it and vice versa.

Polyacrylamide gels are prepared between two glass plates that are held apart by spacers. These gels are run in a vertical position in contrast to the horizontal agarose gels. Polyacrylamide gels can be used to separate DNA as well as proteins.

On the other hand, in SDS-PAGE, denaturing polyacrylamide gels are polymerized in the presence of a denaturing agent (urea, formamide or SDS). A denaturing agent like SDS makes all proteins negatively charged so that under the influence of an electric field, movement occurs only due to mass of the protein and not due to the charge its amino acids contain.

Buffer systems used in vertical gel electrophoresis systems can be continuous or discontinuous. For a much greater resolution, a discontinuous system is preferred. A discontinuous system, as the name suggests consists of a non-restrictive, large pore stacking gel layered on top of a small pore, resolving (separating) gel. The pH of the stacking buffer is 6.8, whereas the pH of the resolving buffer is 8.8.

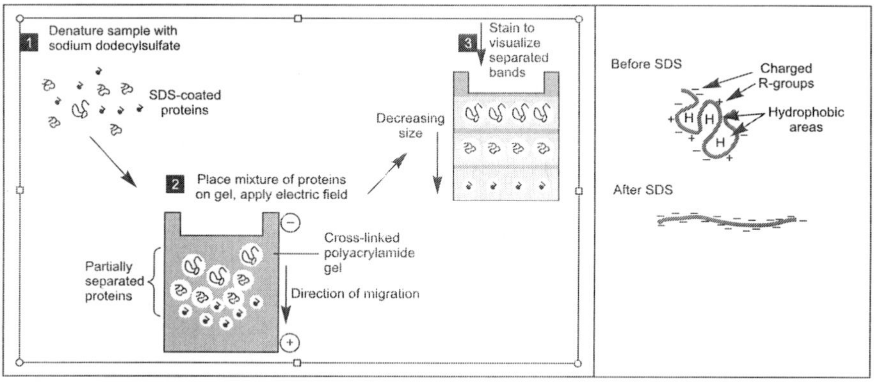

Fig. 7.1 Basic principle behind SDS PAGE. 1: Denaturing of proteins using sodium dodecylsulfate. 2: Basic SDS-PAGE apparatus showing the migration of proteins from cathode to anode. 3: Staining of gels to visualize the bands.

Source: Lodish, H., A. Berk, S. Lawrence Zipursky, P. Matsudaira, D. Baltimore, J. Darnell. 1999.

PROCEDURE

1. Streaked *E. coli* DH5α (or any other bacteria whose protein needs to be isolated) from glycerol on LB agar plate and keept at 37°C overnight.
2. Picked up a single colony with a platinum loop and inoculated it in 5 mL LB medium in a test tube. Kept overnight at 37°C with constant shaking. This is the primary inoculum.
3. Inoculated 400 mL LB with 1% primary inoculum i.e., 4 mL (1 mL primary inoculum for every 100 mL of the medium). This is known as secondary inoculation. Kept at 37°C with constant shaking till $O.D._{600}$ reaches 0.6.
4. Harvested the cells by centrifugation at 10,000x *g* for 10 min at 4°C.
5. Resuspended the pellet in 15 mL of 40 mM imidazole, 4 M NaCl, 160 mM Tris-HCl, pH 7.9.

6. Sonicated the resuspended cells for 10 cycles of 30 s each with an interval of 30s between subsequent sonication. Sonication must be performed on ice.
7. Pelleted down the debris by centrifugation at 10,000x g for 10 min/4°C.
8. The supernatant or clear lysate has whole cell proteins and was collected in a separate falcon tube.

(i) Sample preparation

1. Mixed 30 μL sample with 30 μL 2X SDS gel loading buffer.
2. Boiled for 5 min.
3. Cooled the samples at room temperature. Samples are now ready for loading.

(ii) Polyacrylamide gel preparation

1. Polyacrylamide gel apparatus consists of the following components: glass plates, gel assembler, combs, gel chamber, electrophoresis tank, casting frame.
2. Cleaned all components with alcohol before use.
3. Assembled two glass plates on the casting frame and placed them on the gel assembler.
4. Sealed three sides of the glass plates by applying molten 1% agar with a P_{1000} pipette. Agar was solidified after 10 min.
5. Poured water between the plates to check for any leakage. In case of leakage the apparatus should be disassembled and assembled again.
6. Prepared 15 mL of 10% resolving gel by mixing the following components in a falcon tube (Table 7.5).

Table 7.5 Preparation of 10% Resolving Gel

S. No.	*Components*	*Volume to be added*
1.	H_2O	5.9 mL
2.	30% Acrylamide mix	5.0 mL
3.	1.5 M Tris (pH 8.8)	3.8 mL
4.	10% SDS	0.15 mL
5.	10% Ammonium persulfate	0.15 mL
6.	TEMED	0.006 mL

APS and TEMED should be added in the last as addition of these leads to initiation of polymerisation.

7. Immediately after adding TEMED, poured the gel between the glass plates with the help of a P_{1000} pipette. Poured up till two-thirds the length of the glass plates. Allowed it to solidify for 20–30 min.
8. Prepared 3 mL of 5% stacking gel by mixing the following components in a falcon tube (Table 7.6).

Table 7.6 Preparation of 5% Stacking Gel

S. No.	*Components*	*Volume to be added*
1.	H_2O	2.1 mL
2.	30% Acrylamide mix	0.5 mL
3.	1.5 M Tris (pH 6.8)	0.38 mL
4.	10% SDS	0.03 mL
5.	10% Ammonium Persulfate	0.03 mL
6.	TEMED	0.003 mL

Add APS and TEMED at the last and pour on top of the resolving gel upto the brim of the glass plates.

9. Carefully inserted the comb between the plates into the stacking gel.
10. Allowed the gel to solidify for 15-20 min.
11. Removed the glass plates along with the gel from the casting frame and placed it in the electrophoresis tank.
12. Filled the tank with 1X Laemmli buffer (running buffer).
13. Carefully removed the comb from the gel.
14. Loaded 20 µL sample into each well.
15. Connected the electrodes to power supply and run at 90V till the dye front reaches the bottom of the gel (Fig 7.2).

(iii) Staining and destaining

1. Removed gel carefully from the glass plate under water.
2. Placed gel into a shallow container and submerged it in CBBR staining solution. Stained for 15 min on a gel rocker.
3. Removed staining solution and added destaining solution (Fig. 7.3). Destained for 30 min on a gel rocker. Destained the gel again with fresh destaining solution overnight.
4. Took gel picture.

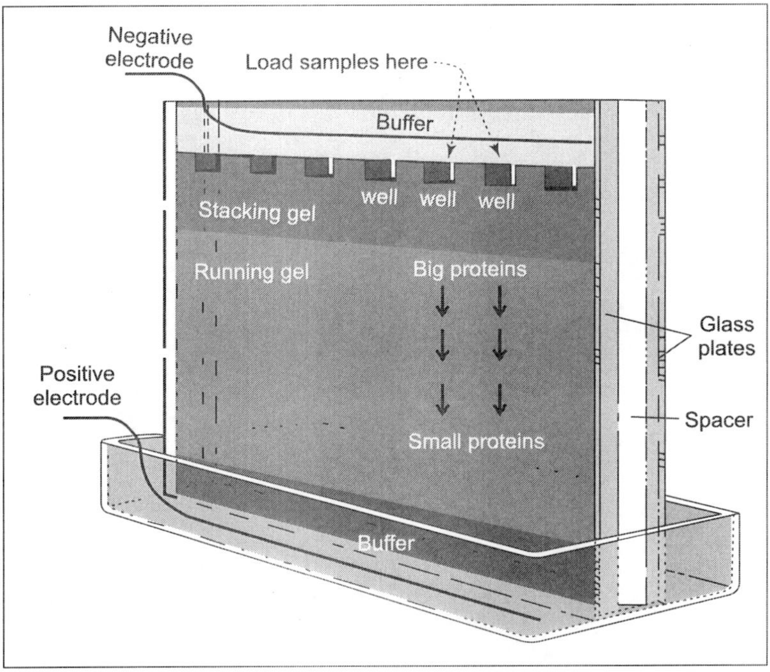

Fig. 7.2 An illustration of the vertical SDS-PAGE apparatus.
Source: https://ww2.chemistry.gatech.edu/~lw26/bCourse_Information/4581/techniques/gel_elect/page_protein.html

Fig. 7.3 Destaining of gel.
Source: Google images

RESULTS

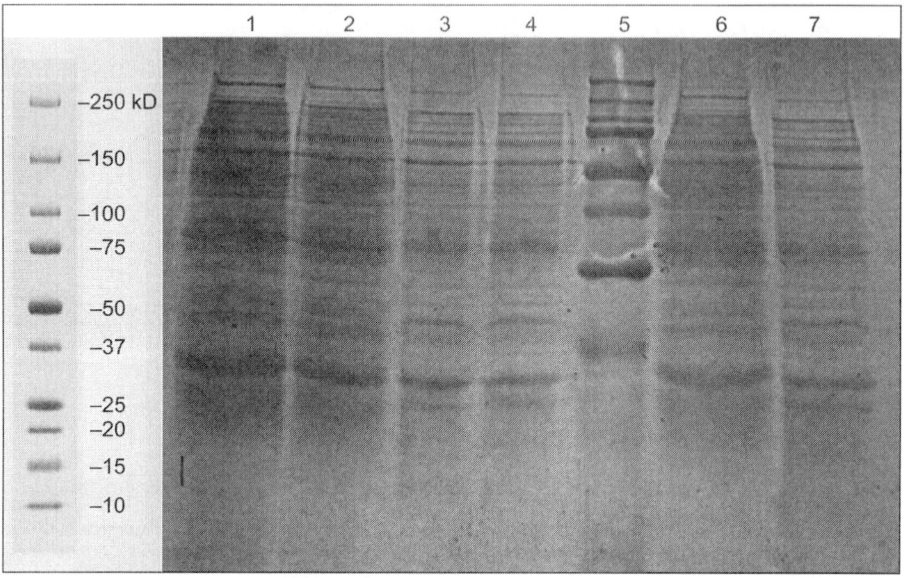

Fig. 7.4 SDS-PAGE gel showing the whole cell protein isolated from a bacterium *E. coli* DH5α. Lane 1-4 & 6-7: Whole cell protein of *E. coli* DH5α. Lane 5: BioRad dual precision marker.

DISCUSSION

The method generates a good profile of whole cell proteins (Fig 7.4). For generation of whole cell protein profiles of different bacteria, the protocol must be modified. Over-expressed proteins can be seen as dark thick bands and under expressed proteins are visible as very thin bands or not at all. This profile is different for different bacteria and is used as a taxonomic marker to differentiate one bacterium from another. Additionally, the mixture of whole cell proteins can be applied to a chromatographic column of choice for purification of a desired target protein. Purified protein thus obtained can be used for further experiments, for example, as enzymes, antibodies, substrates according to the design of the experiment.

FREQUENTLY ASKED QUESTIONS

1. What is the relationship between pore size and migration of protein?

2. What is the role of SDS in PAGE?
3. What is the difference between continuous and discontinuous electrophoresis?

REFERENCES

1. **Brahmachari, S., M. Mukerji, B. Pillai, S. Sen** (2007). Genomics and functional genomics, in *Textbook of Biotechnology* (Editor: Das H. K.) Wiley Dreamtech India. 895-960.
2. **Lodish, H., A. Berk, S. Lawrence Zipursky, P. Matsudaira, D. Baltimore, J. Darnell** (1999). Protein structure and function, in *Molecular Cell Biology* (Editors: Ahr K., Steyn R., DiVittorio S.) W.H. Freeman and Company, New York. 60-95.

8
Quantification of DNA

Nitish Kumar Mahato
E-mail: *nitishyayati.du@gmail.com*

INTRODUCTION

After isolation of DNA, **quantification** and **analysis of quality** are necessary to establish the approximate quantity and purity of the sample. The amount and quality of isolated DNA is crucial to the success of the assays. Within an assay and/or study, conformity of DNA concentration leads to authentic, robust and reproducible results, and assures efficient use of worthwhile samples.

LEARNING OBJECTIVES

To be able to

- Quantify the DNA sample
- Determine the purity of the DNA sample

There are several methods available to quantify a pure DNA sample. The traditional method involves measuring the amount of ultraviolet radiation absorbed by the bases at 260 nm on a spectrophotometer.

QUANTIFICATION OF DNA USING NANODROP

REQUIREMENTS

Glassware and Plasticware

- Microtips

Chemicals

- Autoclaved Milli-Q water or 1X TE
- DNA Sample

Equipment

- Nanodrop ND-1000 spectrophotometer
- Micropipettes

THEORETICAL ASPECT/PRINCIPLE

Spectroscopy is one of the most commonly used technique to identify and quantify solutions of biological macromolecules by characteristic absorption spectrum. The aromatic ring structure of purine and pyrimidine moieties composing the nucleoside bases of DNA and RNA are responsible for absorption of UV light at 260 nm. Nucleic acids as a macromolecule absorb maximally very near to 260 nm. More light absorbed by the sample, higher the nucleic acid concentration in the sample. Estimation of nucleic acid purity is done by determining the ratios of spectrophotometric absorbance at 260/230 and 260/280. Pure preparation of DNA have 260/280 ratio of ~1.8, while for RNA the ratio is ~2.0. Low ratios can be due to protein or phenol contamination in the prepared DNA or RNA sample.

Its principle is based on the Beer-Lambert's law, which is a combo of two laws each dealing with absorption of light related to concentration of absorber and path length or thickness of layer, provided an absorbing substance is partially transparent it will transmit a proportion of the radiation incident upon it. Ratio of intensities of transmitted and incident light presents the transmittance T.

Transmittance (T) = I / I^o

where I^o = Intensity of incident radiation

I = Intensity of transmitted radiation

Intensity of transmitted light **I** is less than **I^o** and debt in the intensity is essentially due to absorption by particles in solutions (apart from the scattering and reflection at the interfaces). This principle is used to make the spectrophotometric instruments.

The Nanodrop ND–1000 is a full spectrum (220–750 nm) spectrophotometer which measures 0.5–1.0 µL of DNA sample with high accuracy and reproducibility. It exploits a patented sample retention technology that uses surface tension alone to hold the sample in place. This eliminates the need for cumbersome cuvettes and other sample containment devices as well as allows clean-up in seconds. In addition, it is capable of measuring samples without dilutions.

PROCEDURE

1. Nanodrop spectrophotometer ND-1000 was switched on and the PC was connected to it.
2. To operate the software
   (a) Icon 'ND-1000' was clicked.
   (b) Nucleic acid was selected from menu.
   (c) Two µL of Milli-Q water or 1X TE (depending upon in which DNA was dissolved) was loaded on to the fiber optic cable and clicked OK to initialize the instrument.

(d) The pedestals were wiped with Kim wipe.

(e) Again 2 μL of Milli-Q water or 1X TE was loaded on to the optic fiber cable. (Depending upon whether DNA sample is dissolved in Milli-Q water or 1X TE.)

(f) The icon Blank was clicked.

(g) Next the sample was loaded (genomic DNA or plasmid DNA) and the icon 'Measure' was clicked.

(h) The concentration of the sample is shown along with its purity.

(i) The sample can be recovered if precious, or pedestals were cleaned using wipes.

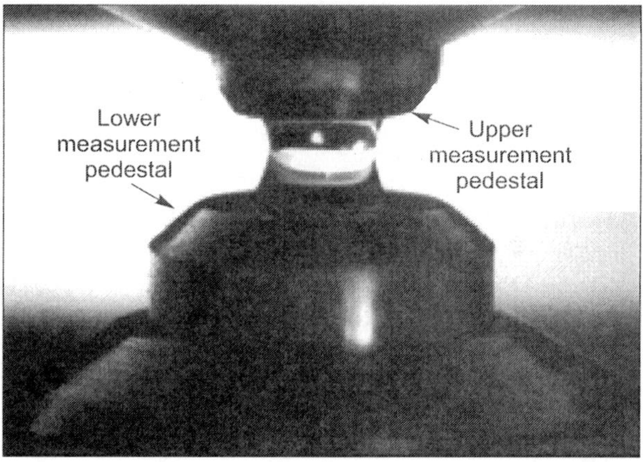

Fig. 8.1 Pedestals position in Nanodrop ND-1000.

OBSERVATION

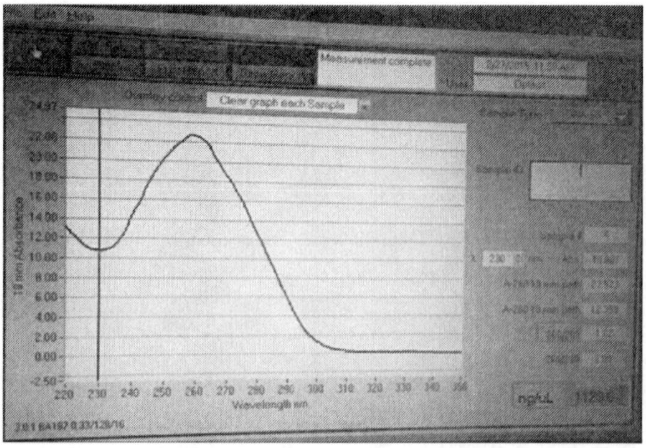

Fig. 8.2 Snapshot of a DNA quantification result shown on Nanodrop ND-1000. Purity and concentration of genomic DNA sample is being shown in right lowermost box. Graph peak shows the maximum absorbance at 260 nm.

RESULT AND DISCUSSION

In Fig. 8.2, quantification result of a genomic DNA sample using Nanodrop ND-1000 is being shown. Concentration of genomic DNA sample is 1128.6 ng/µL and $A_{260/280}$ ratio is 1.82. Peak of the graph also confirms maximum absorbance of the genomic DNA sample at 260 nm. Thus, it can be inferred that quality and quantity of DNA sample is satisfactory and can this sample can be used in further assays.

DRAWBACKS

- Double stranded DNA cannot be distinguished from single stranded DNA by this method.
- DNA quantification using Nanodrop ND-1000 does not reproduce result if experiment is repeated.

QUANTIFICATION OF DNA USING AGAROSE GELS

Requirements

Glassware and Plasticware

- Micropipette
- 1.5 mL microfuge tubes

Chemicals

- 6X loading dye
- Agarose
- 0.5X TBE buffer
- Ethidium bromide

Equipment

- Electrophoresis box
- Gel tray and comb
- UV transilluminator/camera

Biological Material

- Lambda DNA with known concentration
- DNA (genomic/plasmid) sample to be quantified

THEORETICAL ASPECT/PRINCIPLE

Gel quantification is a technique to get an estimation of quantity of DNA bands in agarose gels. The technique basically involves the visual observation of an unknown quantity of DNA to known quantity of DNA on an agarose gel. The gel is stained with ethidium bromide (EtBr) and photographed in a gel documentation system. The brightness of a particular band can be used to estimate the quantity of DNA present in that band when compared to a sample (lambda DNA) of known concentration.

PROCEDURE

1. Lambda DNA of 25 ng/μL concentration was prepared from the stock solution available (0.25 mg/mL).
2. For 25 ng/μL, 3 μL of stock was added to 27 μL of autoclaved Milli-Q water.
3. 0.8% agarose gel (0.8 g agarose in 100 mL 0.5X TBE buffer) was prepared in a flask. 2 μL of 10 mg/mL EtBr solution (wearing gloves since carcinogenic) was added for every 100 mL of agarose and carefully poured (to avoid air bubbles) onto casting tray with the appropriate combs set in place depending on the number of samples.
4. After the gel was solidified (approx. 30 min), combs were removed carefully and tray was set in the electrophoresis tank. 0.5X TBE was used as electrophoresis buffer and added just enough to cover the gel surface.
5. 2 μL, 3 μL, 4 μL and 8 μL of 25 ng/μL lambda DNA (50, 75, 100 and 200 ng respectively) were loaded in the first four consecutive wells which would be served as standards.
6. 1 μL of total genomic DNA was taken and diluted 200X by adding 199 μL of autoclaved Milli-Q water, or 2 μL of plasmid with 2 μL of autoclaved Milli-Q water. DNA was loaded without any dilution, if DNA is expected to come in lower yields (ng amounts). Here, genomic DNA was used in the following amount for quantification: 2, 4 and 10 μL.
7. Electrophoresis was run for 10 min at 100 volts and then for 90 min at 70 volts.
8. After this, DNA bands were visualized under UV transilluminator and intensity of bands of each DNA sample with the standards were compared to estimate it. Gel picture was saved for future reference.

OBSERVATION

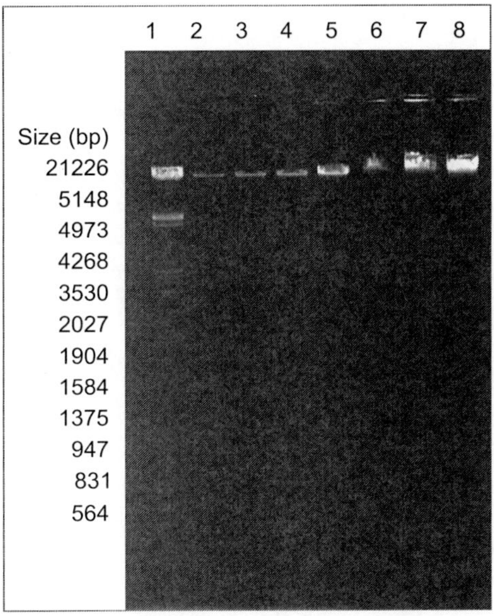

Fig. 8.3 0.8% agarose gel electrophoresis picture showing DNA to be quantified. Lane1: λDNA digested with *Eco*RI and *Hind*III (marker): #SM0191, Fermentas Lane2-5: Lambda DNA standards (50, 75, 100 and 200 ng) respectively, Lane 6-8: Genomic DNA sample of varying amounts.

RESULT AND DISCUSSION

Above gel picture of 0.8% agarose gel electrophoresis (Fig. 8.3) clearly shows varying intensity of bands as quantity of Lambda DNA standards increases. DNA present in the sample can be quantified by comparing the intensity of band with the standards. For example, here, band in the lane 6 is comparable to band of 100 ng lambda DNA in lane 4. So, it can be inferred that almost same amount, i.e., 100 ng of DNA is present in lane 6.

FREQUENTLY ASKED QUESTIONS

1. Why Nanodrop is preferred over spectrophotometer for quantifying DNA?
2. What does $A_{260/280}$ denote in Nanodrop result?
3. Can RNA and proteins be quantified using nanodrop? If yes, discuss the conditions.
4. How can DNA be quantified using agarose gels?

REFERENCES

1. http://www.nanodrop.com/library/nd-1000-v3.7-users-manual-8.5x11.pdf
2. http://vogelchair.wsu.edu/protocols/DNA%20quantification.pdf

9

Primer Designing

Roshan Kumar
E-mail: *roshanzhc@gmail.com*

INTRODUCTION

A primer is a short synthetic oligonucleotide, which serves as a starting point for DNA replication in several molecular biology techniques, such as PCR, DNA sequencing, etc. These primers are designed in a way to have the sequence generally 18–30 nucleotides long which is complementary to the template strand to which our gene of interest belongs. Further, in order to design a proper set of primers, we need an already known template. Mostly, the primers are used for amplifying a gene of interest and gaps in genome assembly followed by sequencing.

To fulfill this objective, one needs to design a primer that should be specific to the gene of interest and can give a high yield. Generally, in PCR reactions, the primers anneal to the complementary region of single stranded molecule, which is then extended by the action of DNA polymerase enzyme. This chapter deals with the characteristic features of a good primer along with the precautions/strategies to be used for primer synthesis.

LEARNING OBJECTIVES

To be able to

- Design a primer for the gene of interest.
- Amplify the 16S rRNA gene/gene of interest from a given bacterial strain.

REQUIREMENTS

- Sequence of gene of interest
- Internet connection

THEORETICAL ASPECT/PRINCIPLE

As we know, a good primer is essential for performing PCR reactions. Following are the important characteristics enlisted for designing a good primer:

1. Melting temperature (Tm) should be in the range of 52–65°C (with G + C content equal to 42–55%).
2. Absence of dimerization ability and should lack the ability of hairpin loop formation.
3. It should lack other priming sites in the template DNA, i.e., the primer should be unique.
4. Low specific binding at 3′-end.
5. The primer should be 18–30 nucleotides in length.

Properties of a Good Primer

Melting Temperature (Tm): Tm is defined as the temperature at which half the DNA strands are single stranded, while other half are double stranded. It is the characteristic of the DNA composition. Higher the GC content, higher will be the Tm due to the presence of triple hydrogen bond between G and C bases.

Calculation of Tm

Shorter than 13 bp

$$Tm = (wA + xT)*2 + (yG + zC)*4$$

Longer than 13 bp

$$Tm = 64.9 + 41*(yG + zC - 16.4)/(wA + xT + yG + zC)$$

where, w, x, y and z refer to the number of A, T, G and C nucleotide, respectively.

- **Annealing Temperature (T_{anneal}):** Annealing temperature is defined as the temperature at which primers anneal to the template DNA. It can be derived from Tm.
- Tanneal = Tm – 4°C
- **Primer Pair Matching:** Primer works in pair, i.e., forward primer and the reverse primer. Both these primers are used in the same reaction, so we should ensure this that the PCR conditions are suitable for both of them.
- **Length:** The length of primer is directly correlated with its uniqueness as well as its melting and annealing temperatures, i.e., the chances of uniqueness increase with the increase in primer

length. Similarly, when the length increases, its melting and annealing temperature also increases. The length of primer should be at least more than 15 bp to ensure its uniqueness. Generally, we consider a primer in the range of 17–28 bp long. But, the range may vary based on its criteria and someone's need.

- **Base Composition:** Generally, random base composition is preferred because the base composition affects melting/annealing temperature and the hybridization specificity. The average G + C content should be around 50–60% that will give us right melting/annealing temperature.
- **Uniqueness:** There should be only one target site for the primer in the template strand, where it can bind. This prevents spurious amplification of the template DNA and helps in amplification of the gene of interest.

  E.g.:-

  5'-ATGAGTGATCTAGACAGACTTGCAAGCCGGGCCGCGATTCAGGACCTCTACTC
  TGACAAGCTCATTGCCGTAGACAAGCGCCAAGAGGGCCGTCTCGCTTCTATTTG
  GTGGGATGATGCAGAGTGGACCATTGAGGGAATCGGCACCTACAAGGGCCCGG
  AAGGCGCCCTCGATTTGGCCAATAACGTACTCTGGCCAATGTTTCACGAATGTAT
  TCATTATGGAACCAATCTGCGCTTGGAATTTGTGAGCGCGGACAAGGTAAATGG
  TATTGGCGACGTCCTTCTCCTTGGAAATCTCGTCGAAGGTAATCAGTCGATTCTT
  ATCGCTGCGGTCTTCACGGATGAGTATGAGCGCCG-3'

  Forward primer CCGCGATTCAGGACCTCTAC

  Reverse primer GCGCTCATACTCATCCGTGA

PROCEDURE

The aim for designing primers can be easily fulfilled by using online primer designing tools, but one should use the tool whose results can be evaluated for a variety of interactions and also specificity for the gene of interest.

Following are some online tools that can be used for primer designing:

- Primer-BLAST
- Primer 3
- Oligo
- PerlPrimer

Steps to design primer using Primer-BLAST

1. The following link http://www.ncbi.nlm.nih.gov/tools/primer-blast/was opened and the sequence was uploaded.

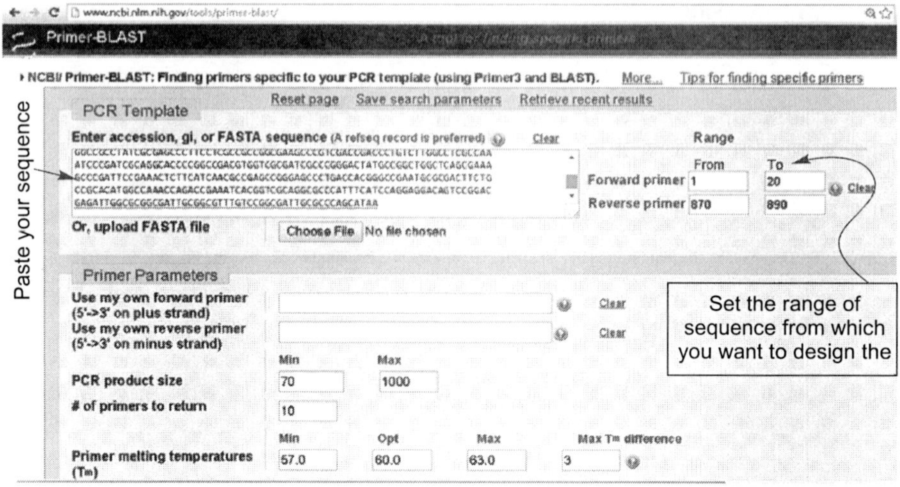

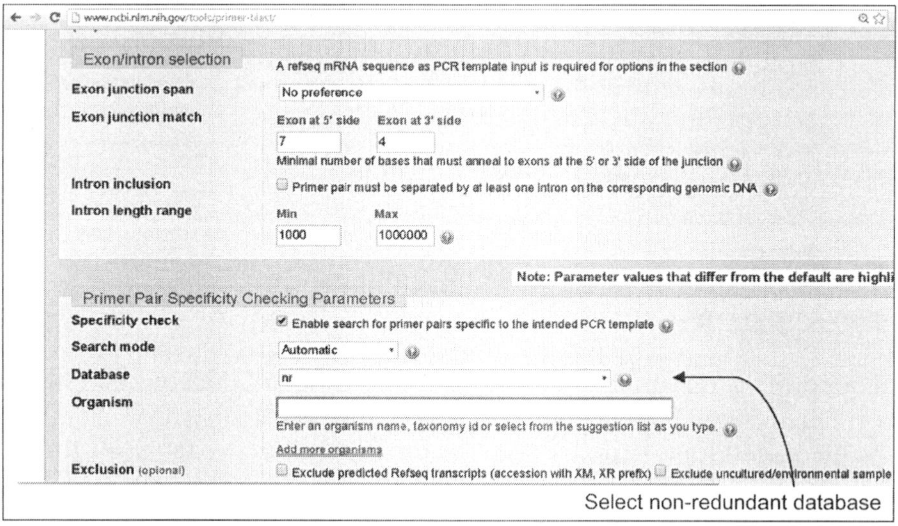

2. 'Get Primer' tab was clicked.
3. A new window opened containing hits of your selected template.
4. The appropriate one was selected and the "Submit" option was clicked.
5. Result will be displayed after some time of the submission.
6. Three primer sets are assigned as default for the input template.
7. It includes forward and reverse primers, sequence of the assigned primer, length, GC content and self-complementarity.

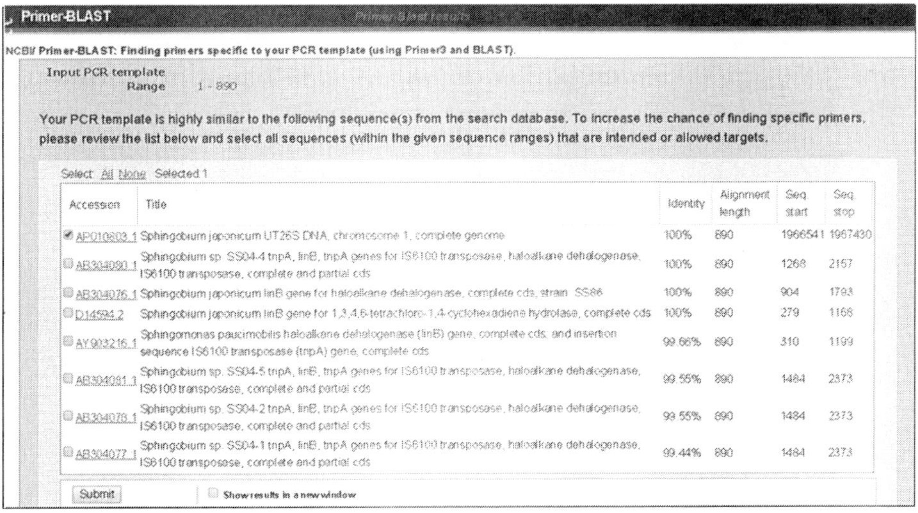

Primer-BLAST

NCBI/ Primer-BLAST: Finding primers specific to your PCR template (using Primer3 and BLAST).

Input PCR template
Range 1 - 890

Your PCR template is highly similar to the following sequence(s) from the search database. To increase the chance of finding specific primers, please review the list below and select all sequences (within the given sequence ranges) that are intended or allowed targets.

Select: All None Selected 1

Accession	Title	Identity	Alignment length	Seq start	Seq stop
AP010803.1	Sphingobium japonicum UT26S DNA, chromosome 1, complete genome	100%	890	1966541	1967430
AB304090.1	Sphingobium sp. SS04-4 tnpA, linB, tnpA genes for IS6100 transposase, haloalkane dehalogenase, IS6100 transposase, complete and partial cds	100%	890	1268	2157
AB304076.1	Sphingobium japonicum linB gene for haloalkane dehalogenase, complete cds, strain SS86	100%	890	904	1793
D14594.2	Sphingobium japonicum linB gene for 1,3,4,6-tetrachloro-1,4-cyclohexadiene hydrolase, complete cds	100%	890	279	1168
AY903216.1	Sphingomonas paucimobilis haloalkane dehalogenase (linB) gene, complete cds; and insertion sequence IS6100 transposase (tnpA) gene, complete cds	99.66%	890	310	1199
AB304081.1	Sphingobium sp. SS04-5 tnpA, linB, tnpA genes for IS6100 transposase, haloalkane dehalogenase, IS6100 transposase, complete and partial cds	99.55%	890	1484	2373
AB304078.1	Sphingobium sp. SS04-2 tnpA, linB, tnpA genes for IS6100 transposase, haloalkane dehalogenase, IS6100 transposase, complete and partial cds	99.55%	890	1484	2373
AB304077.1	Sphingobium sp. SS04-1 tnpA, linB, tnpA genes for IS6100 transposase, haloalkane dehalogenase, IS6100 transposase, complete and partial cds	99.44%	890	1484	2373

Submit   Show results in a new window

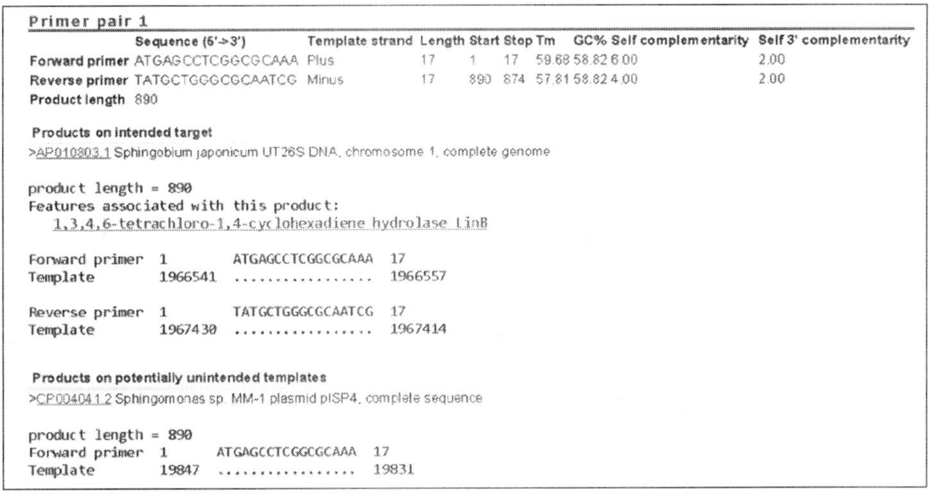

Primer pair 1

	Sequence (5'->3')	Template strand	Length	Start	Stop	Tm	GC%	Self complementarity	Self 3' complementarity
Forward primer	ATGAGCCTCGGCGCAAA	Plus	17	1	17	59.68	58.82	6.00	2.00
Reverse primer	TATGCTGGGCGCAATCG	Minus	17	890	874	57.81	58.82	4.00	2.00
Product length	890								

Products on intended target
>AP010803.1 Sphingobium japonicum UT26S DNA, chromosome 1, complete genome

```
product length = 890
Features associated with this product:
   1,3,4,6-tetrachloro-1,4-cyclohexadiene hydrolase LinB

Forward primer  1         ATGAGCCTCGGCGCAAA  17
Template        1966541   .................  1966557

Reverse primer  1         TATGCTGGGCGCAATCG  17
Template        1967430   .................  1967414
```

Products on potentially unintended templates
>CP004041.2 Sphingomonas sp. MM-1 plasmid pISP4, complete sequence

```
product length = 890
Forward primer  1       ATGAGCCTCGGCGCAAA  17
Template        19847   .................  19831
```

FREQUENTLY ASKED QUESTIONS

1. What are the basic characteristics that need to be followed during primer designing?
2. What is the difference between melting temperature (Tm) and annealing temperature (Tanneal)?
3. How is Tm calculated for primer with size less than and more than 13 bps?

REFERENCES

1. **Ye, J., G. Coulouris, I. Zaretskaya, I. Cutcutache, S. Rozen, T. Madden** (2012). Primer-BLAST: A tool to design target-specific primers for polymerase chain reaction. *BMC Bioinformatics* **13:** 134–144.
2. **Kibbe, W.A.** (2007). OligoCalc: an online oligonucleotide properties calculator. *Nucl. Acids Res.* **35**: 43–46.

10

Polymerase Chain Reaction (PCR)

Roshan Kumar
E-mail: *roshanzhc@gmail.com*

INTRODUCTION

Polymerase Chain Reaction (PCR) is one of the milestones in the history of molecular biology. It was developed in the year 1983 by **Kary Mullis** and **Fred Faloona**. Since then this tool has emerged as a backbone in the area of molecular biology with time to time modifications. PCR is based on the ability of DNA polymerase enzyme to synthesize new strands of DNA that are complementary to the template strand added in the reaction mixture. Basically, DNA polymerase adds a starter nucleotide on 3′-OH group of the primer provided in the reaction mixture. This is further extended in order to synthesize a specific region of the template strand again and again. After 35–40 cycles of PCR reaction, the gene of interest is amplified in billions of copies (Fig. 10.1).

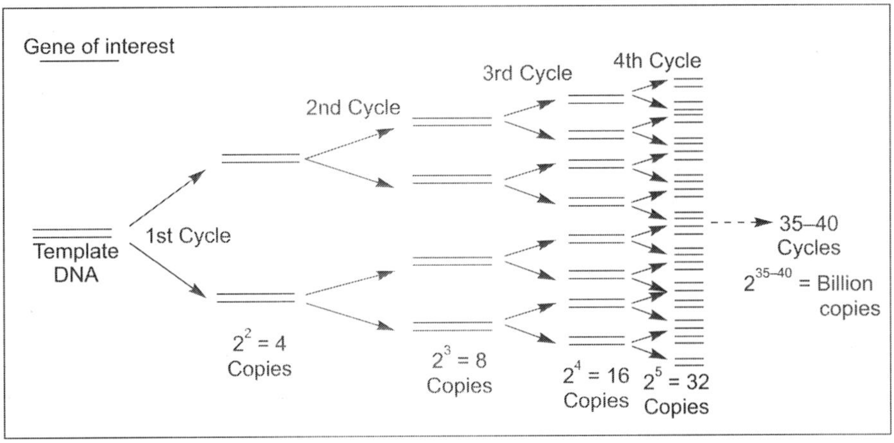

Fig. 10.1 Schematic representation of the amplification by polymerase chain reaction (PCR).

LEARNING OBJECTIVES

To be able to

- Amplify the 16S rRNA gene or any gene of interest from a given bacterial species.

REQUIREMENTS

Glassware and Plasticware

- Microfuge tubes (1.5 mL)
- Microtips
- Disposable gloves
- Gel scoop

Chemicals for Reaction

- DNA
- Taq. Buffer
- dNTPs
- Primers (8F and 1542R)
- DMSO
- Taq DNA polymerase
- Milli-Q water

Chemicals for Running the Gel

- Agarose
- Bromophenol blue (10X stock)
- Ethidium bromide
- 10X TBE, pH 8.0
- Milli-Q water
- 100 bp ladder (Fermantas)

Equipment

- Thermal cycler
- Vortex
- Submarine/horizontal gel electrophoresis apparatus with power supply
- Micropipettes
- UV-transilluminator
- Gel documentation system

Biological Material

- DNA isolated from a bacterial species.

Role of chemicals

Template DNA strand: (prokaryotic/eukaryotic genomic DNA, plasmid/cosmid DNA, etc.)

Primers: Mix the known concentration of the primer into Milli-Q. Two sets of primer which are complementary to the 3′ends of each of the sense and antisense strand of the template DNA.

Buffer (10X): $MgCl_2$ (25 mM is convenient and is usually already present in the buffer.)

The buffer provides a basic chemical environment for the activity and stability of DNA polymerase enzyme. Mg^{2+} plays a crucial role in addition of dNTPs as it forms the active center of polymerase enzyme. For dNTP addition, the polymerase needs to be in a certain orientation with respect to the dNTP. Mg^{2+} helps in proper orientation of dNTPs for nucleophilic attack from the growing oligonucleotide strand.

When concentration of Mg^{2+} is too less, primers fail to anneal to target DNA and at high concentrations, false priming takes place.

Taq polymerase: Taq polymerase or other DNA polymerase which has optimum activity at around 70°C.

Deoxynucleotide triphosphates (dNTPs) (2 mM or 2.5 mM stock): The dNTPS act as building blocks for the synthesis of new DNA strand.

Note: The concentration of dNTPs stock is 2.5 mM which means the final concentration of each dNTP (dATP, dGTP, dCTP, and dTTP) is 2.5 mM – NOT that all dNTPs together make 2.5 mM.

Dimethyl Sulphoxide (DMSO): 3 µL in a 100 µL reaction is enough and is not an obligatory addition to the reaction. It has been shown to facilitate DNA strand separation (in G + C= rich secondary structures) because it disrupts base pairing by disrupting hydrogen bonds in DNA and improves PCR efficiency.

The final working concentration of reagents in PCR reaction is as follows:

Buffer: Working 1X, usually comes as 10X stock. For 25 µL reaction, 2.5 µL of buffer is added. That is 1/10th of the total reaction mixture.

dNTPs: Final concentration to be 200 µM or add 6 µL in 100 µL reaction.

Primers: A considerable primer concentration is 50 pmol of each primer per reaction. If you don't get desired product, concentration can be increased to 75 pmol or 100 pmol. 2.5 µL (of 20 µM stock) in 100 µL reaction is commonly used.

Template: PCR product can be obtained with incredibly smaller amounts of template DNA. If it is a pure DNA then 1–3 ng/µL is enough but in case of impurities, it is important to further dilute it.

16S rRNA Gene Amplification

Reaction Conditions

Initialization step: This initialization step consists of heating the reaction mixture to a temperature of around 94–96°C for 1-9 min. This is required for the activity of DNA polymerase as it needs heat activation.

Denaturation step: This is the first regular step of PCR which consists of heating the reaction mixture to 94–98°C for 20–30 s. It denatures the dsDNA molecule by disrupting the hydrogen bonds that exist between complementary bases and provides ssDNA molecule for further reactions.

Annealing step: The reaction temperature is lowered to 50–65°C for 20–40 s which helps annealing the primers to the single-stranded DNA template. Generally, the annealing temperature (T_{anneal}) is about 3–5°C lower than the Tm of the primers.

Extension/elongation step: At this step, the temperature depends on the DNA polymerase used in the reaction mixture. Taq polymerase which is often used has its optimum activity at 75–80°C and mostly a temperature of 72°C is preferred for the enzyme.

Final elongation: This is the final step of each PCR cycle which takes place at a temperature of 70–74°C for 5–15 min to ensure that any remaining single-stranded DNA is fully extended.

Final hold: This step is basically for short-term storage at 4–15°C.

PREPARATION

Make 0.5X TBE (working concentration) by adding 1 part of 10X TBE to 19 parts of Milli-Q water. 0.8% agarose gel is prepared as explained in chapter "Visualization of Nucleic acids and Proteins".

PROCEDURE

1. For a 100 µL reaction, added all the contents in the PCR vial (enzyme was added at last).
   - DNA: As per purity of the sample (1–2 µL)
   - Primers: 2.5 µL each (8F and 1542R)
   - dNTPs: 6 µL
   - Taq buffer: 10 µL
   - DMSO: 3 µL
   - DNA polymerase: 0.8 µL
   - Milli-Q: to make up volume to 100 µL

2. Contents were mixed well and flash spin was given.
3. The following conditions were set on the thermal cycler for 100 µL of reaction volume (30 cycles)
   - 96°C: 5 min
   - 96°C: 45 s
   - 55°C: 45 s
   - 72°C: 1.30 min
   - 72°C: 7 min
   - 4°C: Hold
4. After the run, samples were loaded on the gel (100 µL+ 3 µL Bromophenol blue dye) and gel was run for around 2h at 80V.
5. The band was cut from the gel using a scalpel by using the transilluminator.

OBSERVATIONS AND RESULT

A band is seen near 1600 bp (16S rRNA gene is near about 1600 bp in length) which can be predicted by using the ladder (100 bp ladder).

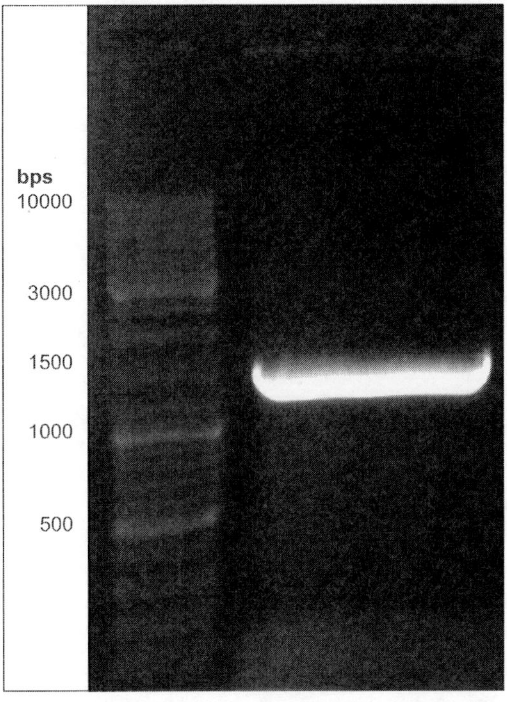

Fig. 10.2 0.8% agarose gel electrophoresis of the amplified 16S rRNA gene product Lane M: 100 bp ladder (Fermentas) Lanes 1-2: Amplified 16S rRNA from bacterial DNA.

DISCUSSION

16S rRNA gene is conserved in sequence among bacterial species, hence it can be sequenced to know the similarity with other species of the same genus and for several other purposes.

Using the polymerase chain reaction the gene was amplified thus increasing the amount and concentration to be used in further experiments.

FREQUENTLY ASKED QUESTIONS

1. Why less reaction volume is preferred while putting PCR reaction?
2. Why normal DNA polymerase cannot be used for PCR reaction?
3. Is the amplification by PCR always error-free?

REFERENCES

1. **Bartlett, J. M., S. D. Stirling** (2003). A short history of the polymerase chain reaction. *PCR Protocols* **226:** 3–6.
2. **Saiki, R., S. Scharf, F. Faloona, K. Mullis,G. Horn, H. Erlich, N. Arnheim** (1985). Enzymatic amplification of beta-globin genomic sequences and restriction site analysis for diagnosis of sickle cell anemia. *Science* **230:** 1350–1354.

11

Restriction Digestion

Rashmi Kumari
E-mail: *rash21289@gmail.com*

INTRODUCTION

Cutting large DNA molecules into smaller pieces using special enzymes is called restriction digestion. These special enzymes are known as restriction endonucleases, exonucleases, molecular scissors (restriction enzymes or REs).

LEARNING OBJECTIVES

To be able to digest DNA with restriction endonucleases.

REQUIREMENTS

Glassware and Plasticware

- Microcentrifuge tube
- Strata cooler
- Dispensers
- Microtips

Chemicals, Buffers and Solutions

Buffers: Restriction enzymes are provided with different buffers. These buffers have appropriate salts and sometimes include essential cofactors that are necessary for the enzyme to function. Major function of the buffer is to maintain pH of the reaction (usually, 8.0) and provide a favorable environment for the enzyme to function. Most companies provide about 4 different kinds of buffers (called A, B, C, D, etc.) and occasionally a "unique buffer" for a particular enzyme which are provided as 10X buffer.

The final concentrations of reagents in restriction digestion are as follows:

Buffer: 1X, usually comes as 10X stock.

For 30 μL reactions, this means 3 μL. (that is, 1/10th of the total reaction mixture).

For Running the Gel

- Agarose
- Bromophenol blue 10X stock
- Ethidium bromide
- 10X TBE, pH 8.0
- Milli-Q water
- Ladder (Fermentas)

Equipment

- Water bath (set at appropriate temperature)
- Electrophoretic system
- UV-transilluminator

Biological Material

- DNA isolated from a bacterial species (plasmid/genomic) **(~100-300 ng).**

THEORETICAL ASPECT/ PRINCIPLE

Restriction enzymes (or restriction endonucleases) are also called "molecular scissors" that cut the double-stranded DNA. These enzymes recognize a specific sequence called restriction site and make two breaks, one through each of the phosphate backbones of the double helix without damaging the bases of the DNA. Although the DNA is broken down into smaller fragments, the chemical bonds can be reformed by another enzyme known as DNA ligases.

Werner Arber, Daniel Nathans and Hamilton Smith in 1978 received the Nobel Prize for their discovery of restriction endonucleases. This discovery led to the development of recombinant DNA technologies. The practical use of restriction enzyme for the first time was done in the field of Science and Medicine in manipulating *E. coli* bacteria that was able to express recombinant human insulin gene for diabetics.

Restriction-modification system is a defensive measure developed in bacteria to protect them from invasion by foreign DNA, basically bacteriophage DNA. Restriction enzymes form an integral part of this restriction-modification system of bacterial cells. In general, there are 4 restriction modification systems present in bacteria that can be differentiated on the basis of restriction site, cofactor requirement, etc. Restriction enzyme cuts DNA segments within a specific nucleotide sequence called restriction sites (Table 11.1). These sequences are generally four, six, eight, ten, or twelve nucleotides long. Recognition sequences may occur by chance in any long sequence. These modification systems can cut the DNA molecule within the recognition site or few nucleotides away from the recognition site. Restriction enzymes may be classified into four types based on their subunit structure, cofactor requirements, specificity of cleavage, and associated methylase activity.

Type I enzymes do not hold much potential to be used in gene manipulation. They exhibit both restriction and DNA modification activities. They are located on different subunits of multifunctional enzyme complexes.

Type II enzymes are the most important class of restriction enzyme and have a number of advantages over other classes of restriction system. Type II enzymes and their corresponding modification methyl transferases act as separate proteins. Restriction and modification are mediated by separate enzymes, which make it possible to cleave DNA molecule in the absence of modification. The enzymes in this class are site-specific, and they hydrolyze specific phosphodiester bonds in both strands of the DNA which lie within or in close proximity to their recognition sequence.

Type III enzymes recognize specific sequences, but cleave 25–27 base pairs away from the recognition site. They require Mg^{2+} as cofactors for the enzyme activity. They are similar to Class I enzyme as they combine restriction and modification activities in a single enzyme complex that are composed of different subunits. They lack both the ATPase activity of Class I enzymes and their absolute requirement for SAM.

Type IV enzymes are similar to type II systems. They have similar cofactors requirement and macromolecular structure as that of class II system. The fact that limits its usefulness in molecular biology is that restriction occurs at a distance from the recognition site.

There are large numbers of restriction enzymes (Table 1) obtained from different bacteria, each of which recognizes unique restriction sites and cuts DNA differently. The recognition sequences usually vary between 4 and 8 nucleotides, and many of them are palindromic, i.e., the base sequence reads the same backwards and forwards.

Example: *Eco*RI recognizes the palindromic sequence GAATTC.

Most commonly-used restriction enzymes are Type II restriction endonuclease. These enzymes may produce:

Blunt End: The enzyme cut both strand of the target DNA at the same point creating blunt ends, e.g., -*Sma*I, *Hae*III, *Alu*I

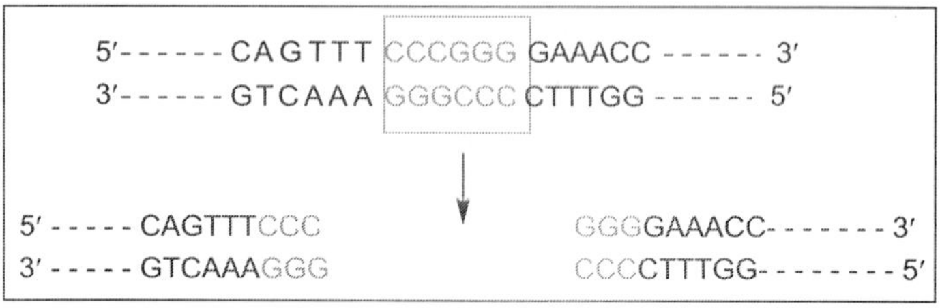

Sticky End: These enzymes cut both strand of the target DNA at different point creating 3′- or 5′-overhangs of 1 to 4 nucleotides so called sticky ends. For e.g., -*Eco*RI, *BamH*I

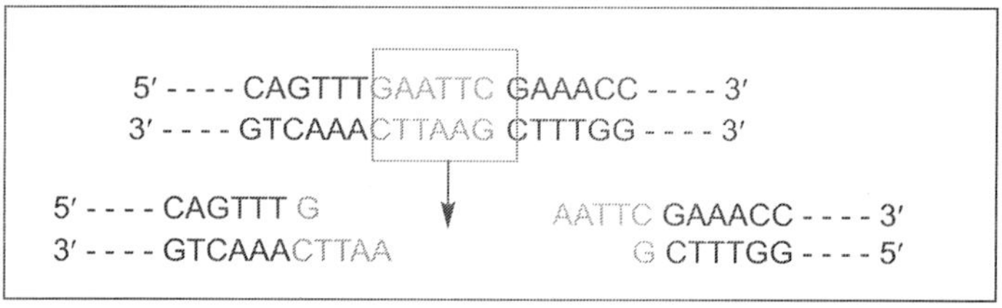

Isoschizomers: They are pairs of restriction enzymes that recognize the same restriction sequence.

e.g., *Sph*I (CGTAC/G) and *Bbu*I (CGTAC/G)

Neoschizomers: They recognize the same sequence, but cut it differently.

e.g., *Sma*I (CCC/GGG) and *Xma*I (C/CCGGG)

Enzyme Activity: One unit of restriction enzyme activity is defined as the amount of enzyme required to produce a complete digest of 1 μg of λDNA in 60 min at 37°C in a 50 μL reaction volume.

Table 11.1 Some Commonly Used Restriction Enzymes

Enzyme	*Source*	*Recognition sequence*	*Cut*
*Eco*RI	*Escherichia coli*	5′GAATTC 3′CTTAAG	5′---G .... AATTC---3′ 3′---CTTAA...........G---5′
*Bam*HI	*Bacillus amyloliquefaciens*	5′GGATCC 3′CCTAGG	5′---G...........GATCC---3′ 3′---CCTAG...........G---5
*Hind*III	*Haemophilus influenzae*	5′AAGCTT 3′TTCGAA	5′---A ....AGCTT---3′ 3′---TTCGA ....A---5′

PROCEDURE

For a **30 µL** reaction,

1. All the contents were added in the 1.5 mL microfuge tube (Table 11.2) (Note: Add enzyme at last).
2. DNA: As per purity and concentration of the sample (100-300 ng)
3. Buffer: 3 µL (10X)
4. Enzyme: [*Eco*RI/*Hind*III (single digestion) or *Eco*RI and *Hind*III (double digestion)]: 1 µL

Table 11.2 Components of Restriction Digestion Reaction Mixture

S. No.	*Components*	*Tube 1* *Hind*III	*Tube 2* *Eco*RI	*Tube 3* *Hind*III + *Eco*RI
1.	Milli-Q water	23 µL	23 µL	22 µL
2.	Lamda DNA (30 ng/µL)	3 µL	3 µL	3 µL
3.	10X Buffer	3 µL (buffer D)	3 µL (unique buffer)	3 µL (buffer D)
4.	Restriction enzyme (20 U/µL)	1 µL	1 µL	1 µL + 1 µL

5. Milli-Q water was added to make up the volume.
6. Keep the sample at 37°C for 1–3 h.

In the meantime agarose gel was prepared for loading the sample.

1. 0.5X TBE was made (working concentration) by adding 0.5 part 10XTBE to 19.5 parts of Milli-Q water 1 part 10X TBE to 19 parts of Milli-Q water and 0.8% agarose gel as described in previous experiments.
2. Gel was allowed to solidify.
3. After this, samples were loaded on the gel (30 µL + 2 µL BPB dye) and the gel was run for around 2h at 80V.
4. Also, an undigested DNA was loaded as a negative control (Figs. 11.1 and 11.2).

RESULT

Lane 1: λ/*Eco*RI+*Hind*III digested ladder (Fermentas)

Lane 2: λ DNA (undigested) (Fermentas)

Lane 3: λ DNA digested with *Eco*RI (6 fragments of which 4 are clearly visible)

Lane 4: λ DNA digested with *Eco*RI and *Hind*III (11 fragments of which 9 are clearly visible here)

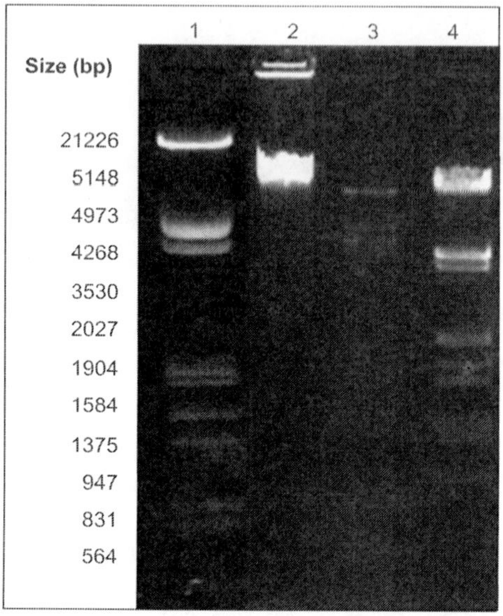

Fig. 11.1 0.8% Agarose gel electrophoresis showing restriction digestion of λ DNA with *Eco*RI and double digestion with *Eco*RI and *Hind*III.

Restriction Digestion of plasmid pRL60, 10.2 kb in size with *Eco*RI, *Hind*III & *Nde*I

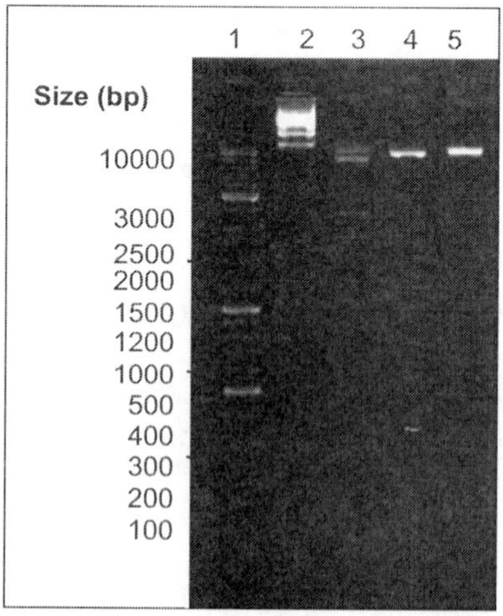

Fig. 11.2 0.8% agarose gel electrophoresis showing restriction digestion of plasmid pRL60 (10.2 kb).

Lane 1: 100 bp ladder (Fermentas)
Lane 2: Undigested plasmid pRL60
Lane 3: Plasmid pRL60 digested wit *Eco*RI (3 fragments)
Lane 4: Plasmid pRL60 digested with *Hind*III (3 fragments)
Lane 5: Plasmid pRL60 digested with *Nde*I (only 1 site so plasmid linearizes)

DISCUSSION

As can be depicted from the gel picture, it shows the digestion pattern of lambda DNA cut with single enzyme *Eco*RI and with double enzymes, *Eco*RI and *Hind*III. For detail of restriction sites, see Fig. 11.3. With the restriction enzymes, the larger DNA fragments have been cut into smaller fragments that are seen in the gel.

Proper restriction digestion depends upon many factors like temperature, concentration of DNA, buffer, etc. However, sometimes the restriction digestion reaction fails due to phenomenon known as "star activity".

This further depends on several factors: absence of Mg^{2+}, high content of glycerol, low salt strength, etc. For example, *Eco*RI under optimal conditions cleaves at G/AATTC, however if star activity is associated, it will cleave at N/AATTN under high pH or low ionic strength, where N represents any nucleotide.

Some factors that inhibit star activity:

- Few units of enzyme should be used to get a complete digestion. This also reduces the final glycerol concentration in the reaction mixture.
- The DNA used for restriction digestion should be pure and free from any organic solvents such as alcohols.
- pH of the reaction buffer should be lowered to pH 7.0.
- Mg^{2+} can be used as the divalent cation.

Another factor associated with improper or partial digestion may involve methylation of DNA. If the target sequence of DNA is methylated, the restriction enzyme will not recognize the site.

Keypoints

- All restriction enzymes should be stored at –20°C.
- Restriction enzyme proteins are stored in glycerol. The enzyme should not be more than 1/10 of final volume of the reaction as 10% concentration of glycerol inhibits digestion and cause star activity. This leads to non-specific and aberrant cleavage of the DNA molecule.

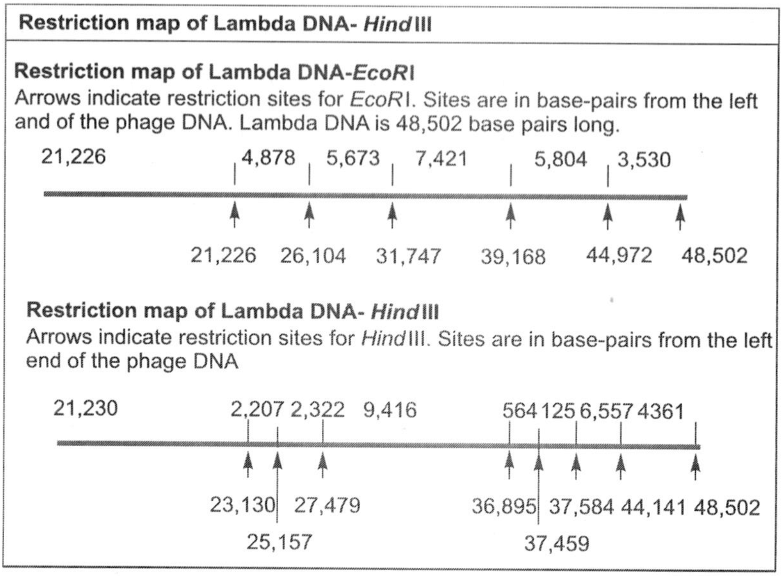

Fig. 11.3 Restriction map of Lambda DNA digested by *EcoR*I (above) and *Hind*III.

- Always set a positive and negative control reaction. This helps to know if the enzyme is working properly and the reaction is correct.
- Compatible buffers should be used, while setting up a double digestion reaction. (Information is available on manufacturer's website.)
- Enzyme and buffer should be kept on ice throughout the process.

FREQUENTLY ASKED QUESTIONS

1. What is the concentration of DNA that should be used for restriction digestion?
2. What are the basic precautions to be used while putting up restriction digestion reaction?
3. What are the conditions for a successful restriction enzyme digest?

REFERENCES

1. http://en.wikipedia.org/wiki/Restriction_enzyme.
2. **Sambrook, J., E.F. Fritsch, T. Maniatis** (1989). Enzymes used in molecular cloning, in *Molecular cloning: A Laboratory Manual*. 2nd Ed. (Editor: Nolan C) Cold Spring Harbor Laboratory Press. 5.1–5.95.

12

Purification of DNA from an Agarose Gel

Priya Singh & Helianthous Verma
E-mail: *priya.impulse@gmail.com*
helianthousverma@gmail.com

INRODUCTION

DNA purification from agarose gel after gel electrophoresis is a necessary step if the amplified product or DNA band is needed for further experiments, such as sequencing, cloning, etc.

LEARNING OBJECTIVES

To be able to

- Excise the bands from the gel using UV-transilluminator.
- Elute the DNA from the excised agarose gel by the kit "Nucleospin Extract II".

Methods to be followed:

- Gel extraction by NucleoSpin® Extract II kit (for small fragments)
- Gel extraction by the enzymatic reaction using enzyme agarase (for high molecular weight DNA).
- Gel extraction from low melting agarose gels also known as the organic extraction method.

1. GEL EXTRACTION BY THE NUCLEOSPIN® EXTRACT II (FOR SMALL FRAGMENTS) KIT

Requirements

Glassware and Plasticware

- Microfuge tubes (1.5 mL)
- Disposable pipette tips

Chemicals

- Nucleospin extract II (all the required chemicals are provided in the kit).
- Milli-Q water to dissolve the DNA

Equipment

- UV-transilluminator
- Scalpel
- Micropipettes
- Water bath
- Centrifuge

Biological Material

Gel slice containing the required DNA fragment.

THEORETICAL ASPECT

DNA binds to a silica membrane in the NucleoSpin® Extract II columns, in the presence of the chaotropic salt in Binding Buffer NT in which the gel slice is melt. This binding mix is loaded onto the NucleoSpin Columns provided in the kit. Salts and soluble macromolecular contaminations are removed by using Wash Buffer NT3 to wash the columns. Bound DNA is then eluted using either Elution Buffer NE provided in the kit (5 mM Tris-HCl pH 8.5) or Milli-Q water.

Kit components:

- The kit NucleoSpin® Extract II is for the purification of PCR products directly and any DNA from TAE/TBE agarose gels.
- The binding buffer constituents are such that it ensures complete removal of any primers from PCR reactions while binding the small DNA fragments and purified with good amount of DNA recovery.
- With this kit DNA fragments from the PCR reactions carried out using buffers with high detergent concentrations can also be purified with good recovery of DNA.
- The prepared DNA stretch can easily be used directly for PCR, automated fluorescent sequencing of the DNA, or any other kind of enzyme reaction.

Note: Buffer NT contains Guanidiniumthiocyanate. It is hazardous if inhaled or if in contact with skin.

Some important points:

- Buffer NE, TE buffer, Milli-Q water or similar low-salt buffer (pH > 7) can be used for the elution of the DNA from the silica column.

Note: pH of the elution solution must be checked before use also EDTA in the TE buffer causes problem in subsequent use. So care must be taken to avoid low recovery of DNA.

- Using 15 µL of the elution buffer 70–95% of the DNA can be eluted in high concentrations. Typically for 50–10,000bp size of DNA fragments.
- For elution, preheat the elution buffer mildly then add required volume of this into column and incubate for 1–2 min. Spin the column along with collecting tube for 1min at 10,000 rpm.
- **Wash Buffer NT3:** To use this **NT3 Concentrate** required amount of the 95–100% ethanol is to be added first. Make mark to label the bottle to indicate that ethanol has been added. It is stable at room temperature for one year at least.

Preparation

- Excised DNA bands from the agarose gel.

Procedure

For PCR clean-up, removal of salts, enzymes, etc., and concentration from samples without the use of SDS following method is suitable:-

1. **Volume of sample** was mixed with **2 volumes** of **binding buffer (NT buffer)** (e.g., for 50 µL PCR reaction and add 100 µL of binding buffer NT and for each 50 mg of agarose gel 100 µL binding buffer is added).
2. A column tube was then placed in a collection tube (2 mL). Sample was loaded into the column tube. Centrifuged for 1 min at 10,000 rpm. Flow-through was discarded.
3. Added 700 µL of the wash buffer (Buffer NT3) to the column. Centrifuged it for 1 min at **10,000 rpm**, the flow-through was discarded and the column was placed back into the collection tube.
4. Centrifuged for **2 min** at **11,000 rpm** to remove wash buffer completely.

   *The spin column should not come in contact of flow-through at any step during the experiment.
5. After the smell of ethanol has gone (that is, the residual ethanol has evaporated, which may inhibit further enzymatic reactions), the column was placed into fresh 1.5 mL microcentrifuge tube. Then added 15–60 µL elution buffer (Buffer NE) or Milli-Q water and left at room temperature for 1 min. Centrifuged at 10,000rpm for 1 min.

2. GEL EXTRACTION BY ENZYMATIC REACTION BY USING AGARASE (FOR HIGH MOLECULAR WEIGHT DNA)

Reqirements

Glassware and Plasticware

- Microfuge tubes (1.5 mL)
- Disposable pippete tips

Chemicals

- Ethanol
- Gel equilibration buffer
  - (i) 10 mM BisTris-Cl
  - (ii) 5 mM EDTA pH-8.0
  - (iii) 0.1 M NaCl
- 5 M NaCl
- Tris equilibrated phenol (pH 8.0)
- Buffer TE (pH-8.0)
- Agarase enzyme

Equipment

- UV-transilluminator
- Scalpel
- Micropipettes
- Water bath
- Microdialysis system and dialysis tubing (optional for large DNA fragments)

Biological Material

Gel slice containing the required DNA fragment.

Theoretical Aspect/Principle

An enzymatic method is used to recover DNA from the gels with low melting agarose (Burmeister and Lehrach 1989). In this technique, the fragment containing DNA is excised from the gel and digested using the enzyme agarase, it hydrolyses the agarose polymer to disaccharide units releasing DNA fragments. This DNA is further purified by the phenol extraction followed by ethanol precipitation. Being very gentle, this method is extremely useful for recovery of DNA with high molecular weight extracted from pulse field gel electrophoresis (PFGE). However, it works well for the recovery of smaller DNA fragments from agarose gel run in constant electric fields.

Procedure

1. Agarose gel was prepared using 1X TAE buffer and agarose gel electrophoresis was performed at 3–6V/cm. UV transillumintor was used for visualization of the gel.

   In TBE, borate ions inhibit ligation reaction and can interfere with the subsequent purification of DNA fragments on glass beads.
2. The slice of gel with DNA of interest was excised and this was then incubated at room temperature for 30 min in the gel equilibration buffer (20 volumes).
3. After cutting, rest of the gel was again visualized so as to confirm that DNA band is removed.
4. Then, the gel slice was transferred into a microcentrifuge tube having equal volume of equilibration buffer as compared to the gel slice.
5. The gel slice was melt for 10 min incubation at 65°C. The solution was cooled to 40°C and then DNase-free enzyme agarase was added (1–2 units agarase per 200 µL of gel slice). The samples were incubated for 60 min at 40°C.

   During this time agarose is digested to oligo- and disaccharides. If desired, DNA solution may be directly used for ligation, restriction digestion, etc., or the DNA can further be purified using following steps.
6. The DNA was purified and concentrated by the method that follows.

To purify small DNA fragments (<20 Kb)

1. DNA was extracted twice with equilibrated phenol.
2. Upper aqueous phase was transferred to fresh tube and then double volume of TE (pH-8.0) was added.
3. Then, added 0.05 volume of 5 M NaCl and 2 volumes of ethanol. The tube was kept for 15 min at 0°C and centrifuged at maximum speed for 15 min at 4°C in microcentrifuge to collect the precipitate.
4. Ethanol was removed and added 500 µL 70% ethanol at room temperature. It was vortexed and centrifuged as above.
5. Supernatant was removed and tube was left at RT with lid open till the smell of ethanol goes off. Now, appropriate amount of TE (pH–8.0) was added and the pellet was dissolved.

To purify large DNA fragments (>20 kb)

1. The sample digested by agarase enzyme was transferred to dialysis bag, which is sealed. Then placed the bag in a beaker 2,000 mL of TE (pH–8.0).
2. The sample was dialyzed/kept in TE for 4–6 hours at 4°C.

Alternatively, the solubilised gel can be dialyzed against TE (pH–8.0) in drop dialysis apparatus (e.g., microdialysis system, Life Technologies).

Large DNA fragments are not subjected to phenol extraction and ethanol precipitation. Adding 30 μM spermine and 70 μM spermidine in dialysis buffer can enhance recovery of large DNA fragments.

3. ORGANIC EXTRACTION: GEL EXTRACTION FROM LOW MELTING TEMPERATURE AGAROSE GELS

Requirements

Glassware and Plasticware

- Microfuge tubes (1.5 mL)
- Disposable pipette tips

Chemicals and Enzymes

- Ammonium acetate 10 M
- Chloroform
- Ethanol
- LMT-elution buffer
- 20 mM tris-HCL
- 1 mM EDTA
- Phenol:Chloroform (1:1, v/v)
- Tris-equilibrated Phenol (pH-8.0)
- 1XTAE buffer
- Buffer TE (pH-8.0)

*Agarose gel is made up of low melting agarose

Equipment

- UV-transilluminator
- Scalpel
- Micropipettes
- Water bath
- Centrifuge

Biological Material

Gel slice containing the required DNA fragment.

THEORETICAL ASPECT

Low melting agarose stands for the agarose that are modified by hydroxyethylation that reduces number of intra-strand hydrogen bonds, melt at low temperature compared to standard agarose. Since they remain as fluid at 37°C so they are suitable to do ligation, restriction digestion, etc. just by adding the portion of gel slice with DNA of interest.

However, the enzymes work less efficiently in presence of melted gels compared to conventional buffers. In current protocol, the gel slice with the DNA of interest is melted, extracted with phenol:chloroform. This method works best for the 0.5 Kb- 5.0 Kb range. Yields outside this range are low.

PROCEDURE

1. The agarose gel was prepared using 1X TAE buffer and the agarose gel electrophoresis was done at 3–6 V/cm. In TBE, borate ions inhibit ligation reaction and can interfere with the subsequent purification of DNA fragments on glass beads.
2. The gel can be visualized using UV transilluminator and with the help of razor and scalpel the DNA bands can be excised.
3. ~5 volumes of LMT buffer were added to agarose slice in microfuge tube. The gel was melted at 65°C for 5 min.
4. The solution was cooled at RT and then equilibrated phenol was added in equal volume. Then it was vortexed for 20s and the aqueous phase was recovered by centrifugation at 4,000 rpm at 20°C for 10 min.
5. The gel was extracted once with phenol:chloroform and once with chloroform.
6. The aqueous phase was transferred to fresh microfuge tube. Then 0.2 volumes of absolute ethanol were added at 4°C. The mixture was stored for 10 min at RT and then the DNA was recovered by centrifugation at 5000x g for 20 min at 4°C.
7. Washed the pellet with 70% ethanol, the excess of ethanol was removed by air drying and the pellet was dissolved in appropriate TE (pH–8.0).

OBSERVATION AND RESULTS

The eluted sample can be checked on gel or nanodrop can be used to determine its concentration and purity (Fig. 12.1).

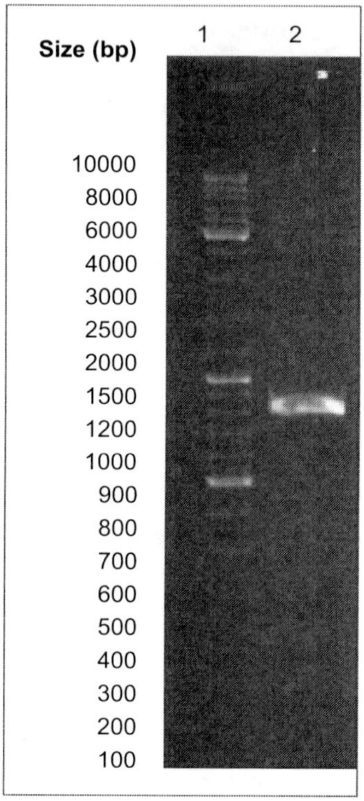

Fig. 12.1 0.8% agarose gel electrophoresis showing the eluted 1.5 kb DNA fragment Lane 1: 100 bp gene ladder (Fermentas); Lane 2: Eluted DNA.

DISCUSSION

Elution of DNA fragments from the agarose gel has been in practice since decades. This includes eluting products from PCR reactions or plasmid isolation. Here, we described three methods for elution of DNA fragments from agarose gel. The first method described here is using NucleoSpin® Extract II kit. This method is suitable for the DNA fragments less than 8Kb and provides high concentration and purified product than the other two methods. The second method, i.e., using agarase enzyme followed by the phenol extraction or dialysis purification are suitable for large DNA fragments. The third method is however based upon the usage of low melting agarose that can further extract using phenol:chloroform and precipitated using ethanol and the protocol works best for the

0.5 Kb–5.0 Kb range. Hence, the usage of these above methods is based upon the product size and its purity. The eluted product should be checked on the agarose gel for its concentration and compared with the marker/ladder. One can also check the concentration by using nanodrop for the last two methods as in the first method; silica membrane can interfere with the nanodrop reading. Presence of the sharp bands at the appropriate position depict successful elution, whereas failed in the same may be due to manual errors while handling the protocol or less initial concentration of DNA.

FREQUENTLY ASKED QUESTIONS

1. Which method would you prefer to elute DNA of size >20 Kb and why?
2. What is the difference between low melting agar and normal agar? When is low melting agar used?
3. What is the best method for checking elution product?
4. What is the principle of working of Nucleo Spin Gel Extraction Kit?

REFERENCES

1. **Sambrook, J., E.F. Fritsch, T. Maniatis.** 1989. Recovery and purification of DNA fractioned on agarose gels in *Molecular Cloning- A Laboratory Manual.* 2nd Edn. (Editor: Chris Nolan) Cold Spring Harbor Laboratory Press. **6:** 22–6.34.
2. **Zumbo, P.** "Phenol-chloroform Extraction". Weill Cornell Medical College (http://physiology.med.cornell.edu/faculty/mason/lab/zumbo/files/PHENOL-CHLOROFORM.pdf)

13

Ligation

Priya Singh
E-mail: *priya.impulse@gmail.com*

INTRODUCTION

Genetic engineering means modification of the genetic makeup of the organisms by addition of foreign traits into its genome and producing new organisms having combination of desired traits. Required gene (coding for desired trait) is taken from the donor genome and ligated into the carrying vector. The recombinant DNA is introduced into the competent organism which would be genetically transformed (modified). This process of genetic modification is carried out effectively by the use of restriction enzymes that are also known as the molecular scissors. They cut the desired DNA and vector very accurately, at the specific restriction sites. Also the ligase enzyme (called molecular glue) is used to paste DNA fragments and the carrying vector. **DNA ligation means joining of the two linear fragments of DNA with covalent bonds.** Ligation proceeds by the establishment of a phosphodiester bond between the 3′-hydroxyl group of DNA strand and 5′- phosphate group of another (Fig. 13.1).

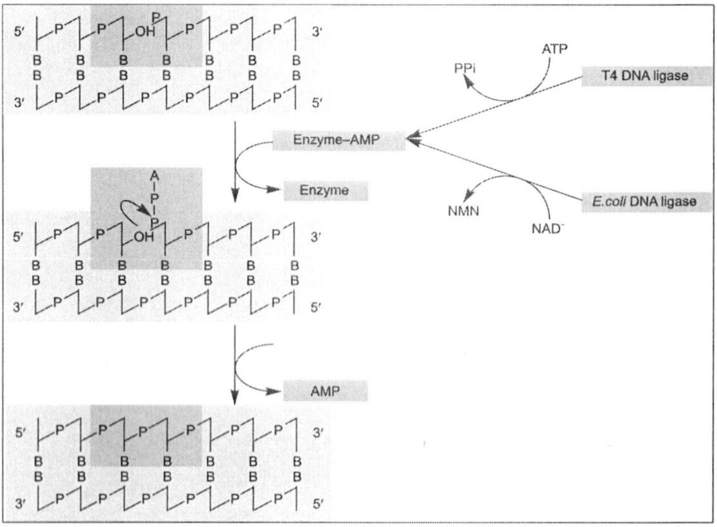

Fig. 13.1 The basic molecular reaction of ligation mechanism.
(*Source:* amrita.vlab.co.in/?sub=3&brch=186&sim=781&cnt=1).

T4 DNA ligase is used for ligation of DNA fragments. It was isolated from the T4 bacteriophage. T4 DNA ligase enzyme ligates DNA fragments having cohesive or overhanging ends. This enzyme is capable of carrying out ligation of fragments with blunt ends as well, but it needs higher concentration of the ligases enzyme for this purpose.

Blunt-end DNA: In such DNA molecules both the strands end in paired nucleotides. These are not good for use in biotechnology because the yield is very low while using a **DNA ligase** to join two such DNA molecules into one.

5′-CTGATCTGACTGAGCTATTGCATCGGT-3′

3′-GACTAGACTGACTCGATAACGTAGCCA-5′

Sticky-end DNA (Overhangs): Such cohesive ends are created by the unpaired overhangs. There is a stretch of unpaired nucleotides in the end of a DNA molecule that forms overhang. These sticky ends can be in any of the two strands, creating either 3′ or 5′-overhang.

5′-GCTGAATG + GATGCGTATGCT-3′

3′-CGACTTACCTACG + CATACGA-5′

Following methods are in use for carrying out ligation of DNA molecules *in vitro*:

- By using the enzyme DNA ligase that covalently ligates the cohesive ends that are produced by certain restriction enzymes.
- Secondly, by using the ability of DNA ligase from phage T4-infected *E. coli* to catalyze the reaction for the formation of phosphodiester bonds between sticky or blunt-ended fragments.
- The last one utilizes the enzyme terminal deoxy-nucleotidyltransferase for the synthesis of homopolymeric 3′-single-stranded tail at the end of DNA fragments.

The common method among all these is the T4 DNA ligase method.

LEARNING OBJECTIVES

- To perform ligation reaction
  1. λ-DNA (size – 48,502 bp)
  2. Plasmid vector and insert DNA

REQUIREMENTS

Glassware and Plasticware

- Microfuge tubes, 1.5 mL
- Microtips

- Disposable gloves
- Gel scoop

Chemicals

For Reaction:

- λ (or any other experimental) digested DNA
- T4 DNA ligase
- Compatible ligase buffers
- Milli-Q water

For Running the Gel:

- Agarose
- Bromophenol blue 10X stock
- Ethidium bromide
- 10X TBE, pH 8.0
- Milli-Q water
- Ladder (Fermentas)

Equipment

- Water bath
- Gel electrophoresis unit
- UV-transilluminator
- Micropipettes

Biological Material

Digested cloning vector DNA and insert DNA (experimental) / λ DNA digested

Role of Chemicals

Digested DNA: *Hind*III digested λ or any other experimental digested DNA is required in order to carry out the ligation of the same with DNA Ligase enzyme. Digested plasmid DNA and insert DNA are required for cloning experiments.

Composition of 10X Ligation Buffer:

- 50 mM Tris-HCl (pH 7.4)
- 10 mM $MgCl_2$
- 20 mM DTT
- 50 μg/mL nuclease free BSA
- 1 mM ATP

Buffer solution used to maintain a suitable chemical environment in reaction tube for the optimum activity and stability of the DNA ligase.

Enzyme: T4 DNA ligase (400U/μL)

1 unit of enzyme: One unit of the ligase enzyme is the amount required to result in 50% ligation of *Hind*III restriction enzyme digested fragments of λ DNA (5′-DNA termini concentration of 300-μg/mL) in 20 μL reaction volume kept for 30 min at 16°C in 1X T4 DNA ligase reaction buffer.

Sterile Milli-Q water

The final concentrations of reagents in restriction digestion are as follows:

Buffer: 1X, usually comes as 10X stock. For 10 μL reactions, this means 1 μL.

That is 1/10th of the total reaction mixture.

DNA: A good concentration of DNA is required for better results (50-100 ng).

PREPARATION

Make **0.5X TBE** (working concentration) by adding 1 part 10X TBE to 19 parts of distilled water and 0.8% agarose gel as described in previous experiments.

PROCEDURE

1. To Perform Ligation Reaction of the λ-DNA using T4 DNA Ligase

For a 10 μL reaction, added all the contents in the PCR vial (add enzyme at last).

DNA: As per purity and concentration of the sample (50–100 ng) - 2 μL of 50 ng/μL of *Hind*III digested λ DNA.

Table 13.1 Components of the Ligation Reaction Mixture

S. No.	*Components*	*10 μL reaction*
1.	10X T4 DNA ligase Buffer	1 μL
2.	Digested λ-DNA	1 μL
3.	T4 DNA Ligase enzyme	1 μL
4.	Milli-Q water	7 μL

The sample was kept at 16°C overnight/3h at room temperature.

After this the samples were loaded on the gel (10 μL+2 μL BPB dye) and the gel electrophoresis was done for around 2h at 80V. Also digested DNA was loaded as negative control.

Observations and result

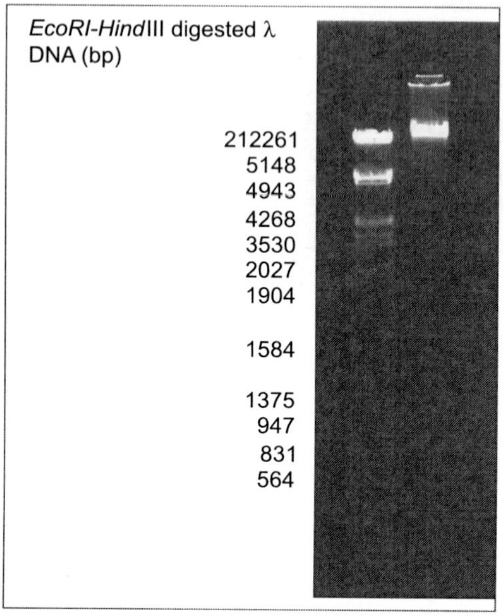

Fig. 13.2 0.8% agarose gel electrophoresis showing the λ-DNA. Lane 1: *Eco*RI-*Hind*III restriction enzyme digested λ-DNA ladder (Fermentas). Lane 2: Ligated λ-DNA.

The ligated λ-DNA is observed as a single DNA band when observed on the agarose gel.

2. Ligation of the Plasmid Vector and Foreign DNA

Ligation of a stretch of foreign DNA to a linearized plasmid vector involves the formation of new bonds between phosphate residues located at the 5′-end of the double stranded DNA and adjacent 3′-hydroxyl moieties. In case both strands of the plasmid vector carry 5′-phosphate residues then four new phosphodiester bonds are generated.

While using dephosphorylated plasmid vectors only two new phosphodiester bonds will be formed. The resultant hybrid molecules possess two single-strand nicks that are repaired after the hybrid DNA has been transformed into the competent bacteria.

Ligation of cohesive termini

Digest the plasmid and foreign DNA's by using the suitable restriction enzymes. If required then the desired fragments of DNA can be isolated by separation using gel electrophoresis. Plasmid DNA can also be treated with phosphatase. DNA is purified by extraction with phenol-chloroform and then ethanol precipitation. Dissolve the DNA in TE (pH 7.6) at final concentration of 100 μg/mL.

Materials required for the dephosphorylation of the vector DNA

- Shrimp Alkaline Phosphatase (SAP) or Calf intestinal Alkaline Phosphatase (CIP)
- 10X buffer
  10 mM $ZnCl_2$
  10 mM $MgCl_2$
  100 mM Tris.Cl (pH 8.3)
- Milli-Q water

Biological material:
Plasmid DNA digested by restriction enzyme.

Procedure

For 100 μL reaction mix the following in mentioned concentrations.

Table 13.2 Components of Dephosphorylation Reaction Mixture

S. No.	*Component*	*For 100 μL reaction*
1.	Digested Plasmid DNA	30 μL(~100ng)
2.	Enzyme: Alkaline Phosphatase	2 μL
3.	Buffer	10 μL
4.	Milli-Q water	58 μL

Steps followed:

The eppendorf carrying reaction mix was kept in water bath set at 37°C for 2 h.

Then this is used as vector DNA for carrying out further ligation.

Phosphorylation of the insert DNA: It is used to add a phosphate group to the 5′-end of the single or double-stranded DNA molecule.

Materials required for phosphorylation of the insert (foreign) DNA

- Insert DNA
- Enzyme – T4 polynucleotide kinase
- 10X PNK buffer
- 10 mM ATPs

Procedure

1. For 30 μL reaction, the following components were mixed in mentioned concentrations.

Table 13.3 Components of the Phosphorylation Reaction Mixture

S. No.	*Component*	*For 30 μL reaction*	*Stock*	*Working*
1.	Insert DNA	20 μL		
2.	Enzyme: T4 polynucleotide kinase	2 μL		
3.	Buffer	3 μL	10X	1X
4.	ATP	3 μL	10 mM	1 mM

2. The eppendorf tube carrying reaction mix was placed in water bath set at 37°C for 30 min.
3. The reaction was stopped by adding 2 μL 0.5M EDTA.
4. Then it was kept again in water bath set at 70°C for 10 min.
5. A flash spin was given and kept at – 20°C until used.

Materials required for ligation of the insert DNA and vector DNA

- PNK treated insert DNA
- Restriction endonuclease digested and SAP treated vector DNA
- T4 DNA ligase enzyme
- 10X buffer

Calculation of the insert amount:

Insert mass in ng =

$$\left(\frac{\text{ng of vector} \times \text{kb size of insert}}{\text{kb size of vector}}\right) \times \text{molar ratio of} \left(\frac{\text{insert}}{\text{vector}}\right)$$

1. Digested plasmid DNA after PNK treatment and Insert DNA after SAP treatment was used further to put up ligation reaction.
2. The ligation reaction was set up as follows:
   (a) 0.1 μg of the vector DNA was transferred to a sterile microfuge tube. An equimolar amount of insert (foreign) DNA.
   (b) 7.5 μL Milli-Q water was added, and warmed the solution at 45°C for 5 min to denature any cohesive termini that might have re-annealed. Then cooled the mixture to 0°C.
   (c) Added:
   - 10X bacteriophage T4 DNA ligase buffer – 1 μL
   - 5 mM ATP – 1 μL

- Enzyme: T4 DNA ligase — 0.1 unit (Weiss unit)

(d) The reaction mix was incubated for 4 h to overnight at 16°C.

***Note:** All these values change in accordance with the concentration of the Insert and Vector DNA.

Two additional control reactions were also set up that contained (1) the plasmid vector alone and (2) the fragment of foreign DNA.

If insufficient foreign DNA is available, use 50–100 ng of the plasmid DNA per ligation and as much as foreign DNA as possible without increasing the volume of the ligation reaction to greater than 10 μL.

1–2 μL of each of the ligation reactions was used to transform competent *E. coli* as described in further chapters of the manual.

Ligation of the Blunt-End DNA

Bacteriophage T4 DNA ligase can catalyse the ligation of the blunt-ended fragments of the DNA, unlike *E. coli* DNA ligase enzyme. Since DNA can easily be made blunt-end, this is an extremely useful property because it allows any DNA molecule to be ligated to any other. However, ligation of blunt-ended termini is a comparatively inefficient reaction and requires four conditions:

- Low concentrations (0.5 mM) of ATP
- The absence of polyamines such as spermidine
- Very high concentration of the ligase enzyme (50 Weiss units/mL)
- High concentrations of blunt-end DNA
- The transformed colonies can be checked by blue white screening method that would be described in subsequent chapters.

FREQUENTLY ASKED QUESTIONS

1. What is meant by 1 unit of T4 DNA Ligase enzyme?
2. What is the purpose of using SAP (Shrimp Alkaline Phosphatase) enzyme to phosphorylate the Vector DNA?
3. State the differences that are there in putting up ligation reaction of cohesive ended and blunt-ended DNA.
4. What should be the control reaction for putting any ligation reaction?

REFERENCE

1. **Sambrook, J., E.F. Fritsch, T. Maniatis** (1989). Strategies for Ligation in *Molecular Cloning - A Laboratory Manual* 2nd Edn. (Editor: Nolan, C.) Cold Spring Harbor Laboratory Press. 1.53–1.72.

Transformation

Anjali Saxena & Namita Nayyar
E-mail: *anjalisaxena1403@gmail.com*
nayyarnamita@gmail.com

INTRODUCTION

Transformation is the most widely and versatile technique used in molecular biology. It is a process of uptake of foreign DNA by bacterial cell.

LEARNING OBJECTIVES

To be able to

- Prepare competent cells (electrocompetent + rubidium chloride)
- Use benchtop cold centrifuge
- Perform transformation by way of **Heat shock method** and **Electroporation**
- Plating of the electroporated/transformed sample
- Selection of clones on the appropriate antibiotic pressure
- Learn how to calculate transformation efficiency

REQUIREMENTS

Glassware and Plasticware

- Conical flasks
- Centrifuge tubes
- Balance
- Microcentrifuge tubes (chilled)
- Cryobox (to store competent cells)
- Electroporator cuvettes
- Dispensers
- Petri plates
- Glass spreaders

Chemicals, Buffers and Solutions

- TFB1- 30 mM potassium acetate; 10 mM $CaCl_2$; 50 mM $MnCl_2$; 100 mM $RbCl_2$

- TFB2- 100 mM MOPS; 75 mM $CaCl_2$; 10 mM RbCl; 15% glycerol
- Glycerol (10% v/v, pre-chilled)
- Milli-Q water (freshly autoclaved and pre-chilled)
- Luria broth for initial growth of culture (5 mL tubes and 100 mL flask)
- LB agar plates with appropriate antibiotic pressure [e.g., ampicillin (working = 150 μg/mL)]

Equipment

- Centrifuge with rotor (pre-cooled at 4°C)
- Ice bucket with ice
- Electroporation device
- Electroporation cuvettes (rinsed with Milli-Q water, UV treated and pre-chilled) fitted with electrodes spaced 0.1–0.2 mm apart
- Incubator shaker (maintained at 37°C).
- Water bath (set at 42°C)

Biological Material

- Plasmid DNA or recombinant DNA (concentration = 50–100 ng/μL)

PREPARATION

1. **10% glycerol (for electrocompetent cells)**
   To 10 mL of glycerol, 90 mL of distilled water was added, autoclaved and stored at 4°C.
2. **LB agar plates with antibiotic pressure of antibiotic (e.g., ampicillin)**
3. **TFB1 (for heat shock competent cells)**
   All the TFB1 components were added in 35 mL water. pH was adjusted to 5.8 with 1 M acetic acid. Volume was made up to 50 mL. Filtered, sterilized (0.2 μM) and stored at room temperature.
4. **TFB2 (for heat shock competent cells)**
   All the TFB2 components were added in 35 mL water. pH was adjusted to 6.5 with 1 M KOH. Filtered, sterilized (0.2 μM) and stored at room temperature.

THEORETICAL ASPECT

Nucleic acids do not enter the cells under their own power; they require assistance in crossing physical barriers at the cell boundary and in reaching an intracellular site where they can be expressed or replicated.

Treatment with calcium ions, or giving electrical current to the cells are the standard methods of making a bacterial cell permeable to DNA. A commonly used technique to introduce plasmids or other DNA fragments into the bacteria is chemically induced competence followed by transformation. On the contrary, during electroporation, the exposure of cells to an electrical discharge reversibly destabilizes their membranes and transiently induces the formation of aqueous pathways or membrane pores that potentiate the entry of DNA molecules. This method of electroporation is comparatively rapid, simple and efficient. In general, the preparation of electro-competent bacterial cells is considerably easier than preparing cells for transformation by chemical methods.

PRINCIPLE

The uptake of DNA from environment by bacterial cell requires the recipient cells to be in specialized physiological state called competence. Competence is distinguished as natural competence and artificial competence.

Natural competence: Frederick Griffith was the first to discover natural competence in 1928. Foreign DNA can be taken up by means of

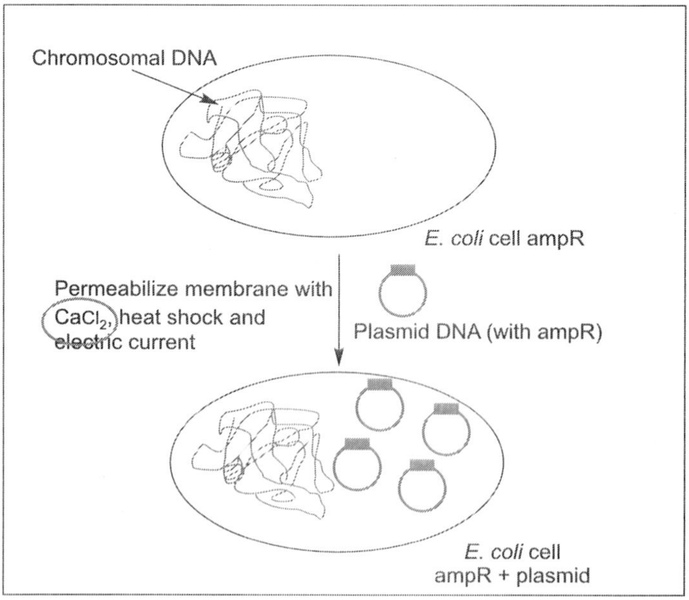

Fig. 14.1 Diagrammatic representation of transformation in *E. coli* with plasmid DNA with ampicillin resistance.

conjugation, transformation and transduction. DNA directly enters into the cell by means of transformation. It is a highly regulated process in bacteria.

Artificial competence: It is not governed by cell's own machinery. Instead, influenced under laboratory conditions, the cells are made permeable to uptake of DNA, by either chemical treatment or by applying electric field. Competent cells can be prepared using the following two methods: Calcium chloride/Rubidium chloride method and electroporation.

In the calcium chloride/rubidium chloride method, bacteria (e.g., *E. coli*), which are not naturally transformable, are grown to mid-log phase, chilled, centrifuged and divalent and multivalent cations are used to induce the cells. Introduction of competency in bacterial cell alters the cell permeability and allows the DNA to cross the bacterial cell membrane. The negative charges on the cell membrane are repelled by negatively charged DNA. In addition, $CaCl_2$ neutralizes the unfavorable interaction between the DNA and the polyanions of the outer layer. The DNA and competent cells are incubated on ice for some time for the stabilization of lipid membrane and allows the interaction of calcium ions and the negative component of cells. Incubation followed by heat shock for 45s, 1 min at 42°C, alters the fluidity of the cell membrane and thereby allowing the DNA molecules to enter the cell through the zone of adhesion.

On the other hand, while preparing the electro-competent cells, the cells are given washes with ice cold Milli- Q water, to reduce the ionic strength of the cell suspension, and then suspended in an ice cold buffer containing 10% glycerol.

The cell suspension can be stored at –70°C for up to 6 months without loss of transformation efficiency. Since *E. coli* cells are small, they require very high field strengths for electroporation. The method outlined in this protocol works well with most strains of *E. coli* and with plasmids <15 kb in size. Very large plasmids would transform *E. coli* with reduced efficiency (Fig. 14.1).

GENERAL CONTROLS TO BE USED DURING THE EXPERIMENT

In every experiment, positive controls should be included, so as to measure the efficiency of transformation, and negative controls are a necessity to eliminate the possibility of contamination and to identify the potential causes of failure.

Negative Control

Negative control should be carried all through the transformation experiment. The negative control comprises an aliquot of competent cells to which no DNA is added. Hence, no transformation should be taking place. The entire aliquot of this negative control should be plated on a single agar plate containing the appropriate antibiotic used to select the transformants. No bacterial colonies should grow on this plate. If any are detected, there can be primarily two possible causes - (1) the competent cells are contaminated with an antibiotic-resistant strain of bacteria during the experiment. This may occur even due to the use of contaminated solutions/reagents during the transformation protocol or (2) defective plates which may have occurred due to addition of antibiotic when the agar was too hot while pouring the plates.

Positive Control

Positive control comprises an aliquot of competent cells which should be transformed with a known amount of a standard plasmid DNA (e.g., pUC18 DNA). The positive control is necessary as it provides a measure of the efficiency of transformation and provides a standard for comparison with previous transformation experiments. In case the positive control does not show transformed colonies, it indicates problems with the competent bacteria or the transformation buffer.

When a ligation reaction is used as a source of transforming DNA, the transformation efficiency is reduced by at least two orders of magnitude than the control because this depends on the amount of recombinant plasmid generated during the ligation reaction.

PROCEDURE

Preparation of Ultracompetent Cells

Important: *All steps in this procedure should be done aseptically in the laminar hood and under cool conditions.*

1. **Primary Inoculation:** A day before preparing the cells, a single colony of *E. coli* DH5α was inoculated from a fresh plate/glycerol stock into 5 mL LB tube and incubated for 18–20 h at 37°C with vigorous agitation (overnight).
2. **Secondary Inoculation:** From the above tube, 100 mL LB flask was inoculated with 1% inoculum (i.e., 1 mL in 100 mL) and incubated for 1½ to 2½ h at 37°C with vigorous shaking at around 200 rpm. For efficient transformation, it is essential that the number of viable cells should not exceed 10^8 cells/mL. To ensure that the

culture does not grow to a higher density, it is advised that the O.D.$_{600}$ should be monitored after regular intervals till it approaches 0.4–0.5.

3. TFB1,TFB2, centrifuge tubes, eppendorfs were ice chilled.
4. 1 mL of overnight culture was inoculated to 100 mL LB flask.
5. O.D.$_{600}$ was monitored until it reaches 0.2 to 0.6.
6. Culture was transferred to centrifuge tubes and chilled on ice for 10–15 min.
7. Cells were harvested at 4000x g for 10 min at 4°C.
8. Supernatant was decanted.
9. Pellet was resuspended in 1/3 original volume chilled TFB1 buffer.
10. Cells were gently mixed and evenly resuspended till no clumps are visible.
11. Cells were incubated on TFB1 on ice for 15 min.
12. Cells were pelleted by centrifugation at 4000x g for 15 min at 4°C.
13. Supernatant was decanted.
14. Cells were resuspended in 1/25th original volume chilled TFB2 buffer.
15. Cells were incubated on TFB2 buffer on ice for 15 min.
16. Aliquots of 100 μL each were transferred into chilled 1.5 mL eppendorf tubes and freezed on ice.
17. Stored at - 80°C.

Preparation of Electro-competent Cells

Important: *All steps in this procedure should be done aseptically in the laminar hood and under cool conditions.*

For preparation of electro-competent cells, the **primary** and **secondary** inoculations were done in a similar manner as for preparation of ultracompetent cells.

1. The flask was chilled on ice (in ice bucket) for ~30 min.
2. Centrifuged at 8000x g/12 min/4°C.
3. Supernatant was decanted and pellet was resuspended in 20 mL of ice cold Milli-Q water.
4. Centrifuged at 8000x g/12 min/4°C.
5. Supernatant was decanted and re-suspended the pellet in 20 mL of ice cold Milli-Q water.
6. Centrifuge at 8000x g/12 min/ 4°C.
7. Supernatant was decanted and re-suspended the pellet in 5 mL of ice cold 10% glycerol (v/v).

8. Centrifuged at 8000x g/12 min/4°C.
9. Supernatant was decanted and re-suspended the pellet in ~200 μL– 500 μL (depending on the size of the pellet) of ice cold 10% glycerol (v/v) to obtain dense slurry of cells.
10. Dispensed the cell suspension into 40 μL aliquots in pre-chilled 1.5 mL microcentrifuge tubes and freezed at -70°C.

TRANSFORMATION

(i) By Heat Shock

1. Competent cells were taken out of - 70°C and thawed on ice (approximately 15–20 min).
2. 1–2 μL of DNA (usually 50–100 ng) was mixed into 50 μL aliquot of competent cells in a microcentrifuge tubes. The bottom of the tube was gently mixed.
3. Competent cells + DNA mixture was placed on ice for 20–30 min.
4. The transformation mixture was heat shocked by placing the bottom of the tube into a 42°C water bath for 1 min 15 s.
5. Tubes were put back on ice for 5 min.

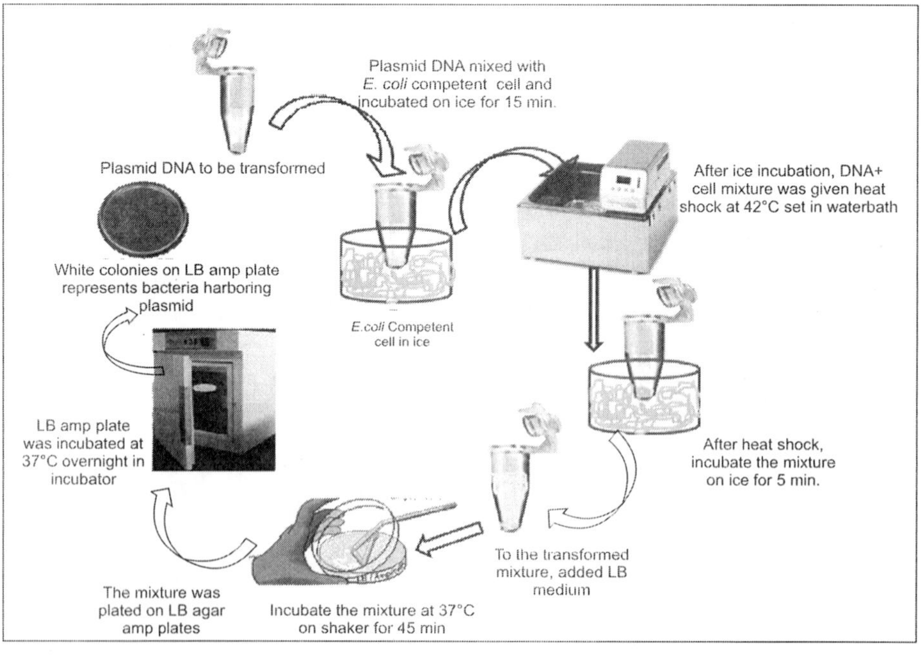

Fig. 14.2 Overview of transformation by heat shock method.

6. 800 μL of LB (without antibiotic) was added and grown in 37°C shaking incubator for 1 h.
7. Some or all of the transformation was plated onto LB agar containing proper antibiotic.
8. Plates were incubated at 37°C overnight.
9. Next day, on the basis of number of colonies obtained, transformation efficiency was calculated (Fig. 14.2).

(ii) By Electroporation

1. Electroporation apparatus was set to deliver an electrical pulse of 25 μF capacitance, 2.5 kV voltage and 200 Ω resistance.
2. To the microfuge tube containing the DNA (50–100 ng in a volume of 1–2 μL only) to be electroporated, 40 μL of electro-competent cells were transferred (thawed on ice) and tubes were kept on ice for 30–60s.

   Note: *When using a DNA ligation reaction mixture for electroporation, either dilute it in Milli-Q or precipitate the DNA followed by its re-suspension in lesser volume of Milli-Q. This helps to prevent arcing upon electric pulse.*
3. DNA/cell mixture was transferred into a pre-chilled electroporation cuvette. Solution was tapped to ensure that the suspension sits at the bottom of the cuvette. Condensation and moisture from the outside of the cuvette was dried and placed it in the electroporation device.
4. An electric pulse was delivered at the above set conditions. A time constant of 4–5 milliseconds should register on the machine.

   Note: *The presence of ions in the electroporation cuvette increases the conductivity of the solution and causes the electric current to arc. Arcing is manifested by the generation of a popping sound in the cuvette during the electric pulse. If arcing occurs in the presence of DNA but not in its absence, remove ions from the DNA preparation by precipitating and re-dissolving it in Milli-Q.*
5. As quickly as possible after the pulse, electroporation cuvette was removed and 500 μL of LB medium was added at room temperature.
6. Cell suspension was added to a 1.5 mL microfuge tube and an additional 500 μL of LB medium was also added making the final volume of 1 mL.
7. Culture was incubated for 45 min to 1 h in a water bath set at 37°C (or alternatively in incubator-shaker) to allow the bacteria

to recover and to express the antibiotic resistance marker encoded by the plasmid (Fig. 14.3).

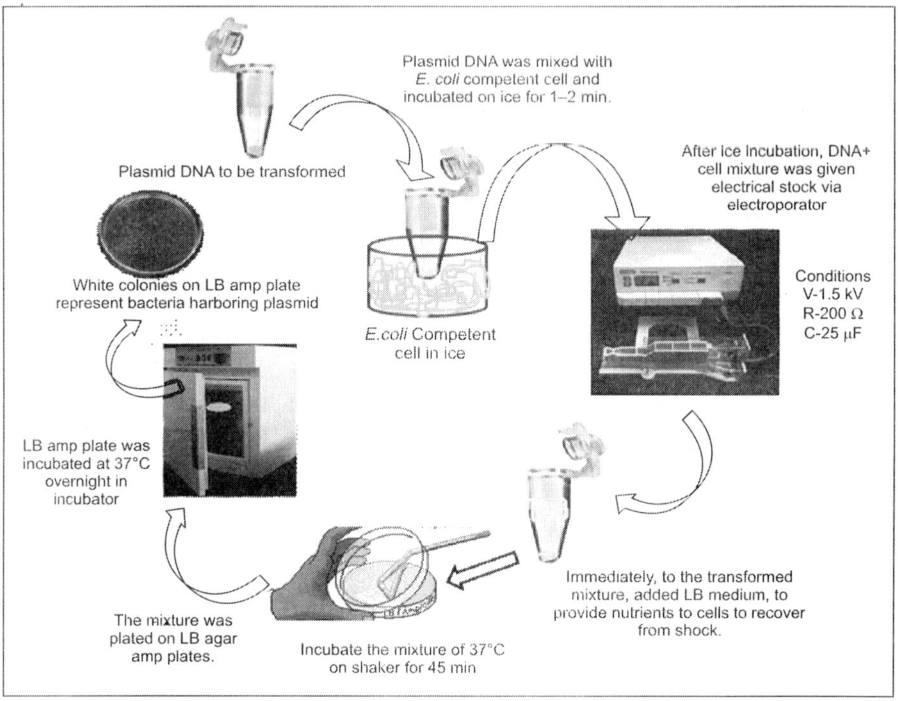

Fig. 14.3 Overview of transformation by electroporation.

8. Different volumes of the electroporated cells were plated onto LB agar plates containing the appropriate antibiotic. Plates were inverted and incubated at 37°C for 12–16 h. The plates should not be incubated for prolonged period (>20 h) at 37°C because the enzyme β-lactamase is secreted into the medium from ampicillin-resistant transformants that can rapidly inactivate the antibiotic in regions surrounding the colonies (Fig. 14.4). This can lead to appearance of satellite colonies.

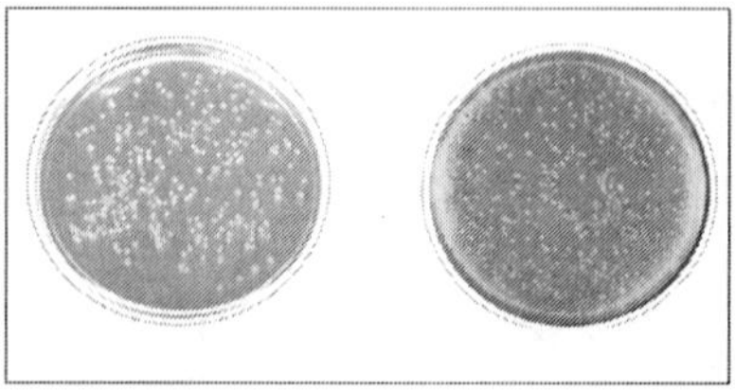

Fig. 14.4 Left panel- White colonies on LB agar amp plates (appeared after transformation of plasmid DNA); Right panel- Blue-white colonies on LB agar X-gal amp plates (appeared after transformation of ligated mixture).

CALCULATION OF TRANSFORMATION EFFICIENCY

Transformation efficiency is defined as the number of transformants obtained per μg of the plasmid DNA used.

Therefore, if you plated 100 μL of the transformed culture and the number of colonies obtained after 12–16 h of incubation is 120, then the calculation to determine the transformation efficiency would be as below:

Number of colonies obtained in 100 μL = 120

Thus, number of colonies in 1 mL = 1200

If, concentration of plasmid DNA used = 75 ng/μL

Total volume of plasmid DNA used = 2 μL

Thus, amount of plasmid DNA = 150×10^{-3} μg

$$\text{Transformation efficiency} = \frac{\text{Total no. colonies}}{\mu\text{g of plasmid DNA}}$$

$$= \frac{1200}{150 \times 10^{-3}}$$

$$= 8 \times 10^{3}$$

FREQUENTLY ASKED QUESTIONS

1. What is the concentration of plasmid DNA that should be used for transformation?
2. What are the basic precautions to be used while preparing competent cells?
3. What are the two types of controls to be used during transformation?
4. What do the blue and white colonies signify?

REFERENCES

1. **Cohen, S. N., C. Chang, L. Hsu** (1972). Nonchromosomal antibiotic resistance in bacteria: genetic transformation of *Escherichia coli* by R-factor DNA. *Proc. Natl. Acad. Sci. U.S.A.* **69:** 2110–2114.
2. **Hanahan, D.** (1983). Studieson transformation of *Escherichia coli* with plasmids. *J. Mol. Biol.* **166:** 557–580.

15

Cloning and Cloning Vectors

Puneet Kohli
E-mail: *puneet.k1311@gmail.com*

INTRODUCTION

Cloning: When a population of genetically identical cells is obtained by growth from a single cell, such a population is called clonal and the process of generating a clonal population is called "cloning". DNA cloning refers to the production of identical copies of a specific DNA sequence generated by the mitotic division of a transformed host cell.

Cloning Vectors: A vector is defined as an autonomously replicating DNA molecule (e.g., plasmid) that has the ability to carry an inserted foreign DNA into a host cell and can proliferate within the host cell so as to produce many copies of itself and the foreign DNA (recombinant DNA).

LEARNING OBJECTIVES

To be able to

- Understand cloning and cloning vectors
- Construct a plasmid map
- Interpret and construct a cloning scheme
- Cloning DNA fragment into the plasmid

REQUIREMENTS

All the requirements are same as mentioned previously in the chapters on plasmid purification, restriction digestions, ligation, transformation and agarose gel electrophoresis.

THEORETICAL ASPECT/PRINCIPLE

Basic essential features of all cloning vectors:

- **Origin of Replication (ORI):** Sequences that permit the autonomous or self-propagation of the vector inside host cell (bacteria or yeast).
- **Small size:** Size lesser than 10 kb is usually considered ideal for a vector. Smaller size ensures ease during genetic manipulation and purification.

- **Multiple Cloning Site (MCS):** Cloning site refers to the region on the plasmid that can be acted upon by many restriction enzymes and thus, a foreign DNA sequence can be inserted into this site.
- **Marker Gene:** Suitable marker genes for identification of the truthful clonal cells (bacteria or yeast) containing the vector along with incorporated foreign DNA.

Plasmids and bacteriophages constitute the major class of commonly used cloning vectors.

I. Plasmids as cloning vectors

(a) Plasmids refer to the self-replicating extrachromosomal circular DNA molecules that are capable of independent existence in bacterial cells and usually confer some unique characteristics to the bacterial cells (e.g., resistance to antibiotics). Thus, plasmids in addition to possessing an **ORI** for replication also harbor selectable marker genes that aid in cloning.

Based on the genes carried by plasmids they are mainly classified as:

1. Fertility (F) plasmids: These plasmids aid in conjugal transfer and carry a set of genes called *tra* genes for the same, e.g., F plasmid of *E.coli*.
2. Resistance (R) plasmids: These plasmids carry set of genes conferring resistance to the bacterium, e.g., resistance against ampicillin.
3. Col plasmids: Plasmids carrying genes coding for collicins.
4. Degradative plasmids: Plasmids carrying genes or degradation of some particular compounds, e.g., toluene.
5. Virulence plasmids: These are also known as the infective plasmids and confer pathogenecity to bacterium, e.g., Ti plasmid of *Agrobacterium tumefaciens*.

(b) **Size** and **copy number** are the two more distinctive properties that make the plasmids an exact candidate for cloning. Their smaller size usually ranging from 1.0 kb to 250.0 kb makes them easy to manipulate. Copy number is dictated by the number of individual plasmid molecules that are found inside the bacterial cells. Based on the size and copy number plasmids are often classified as:

- **Stringent:** This represents the class of plasmids which are larger in size with lower copy number and with an approximate size of 100 to 1000 bp insert limit.
- **Relaxed:** Plasmids with smaller size and usually present in multiple copies within the bacterial cell constitute the relaxed plasmids. These are more valuable for cloning.

Commonly used plasmid cloning vectors: pBR322, pUC18, pUC19

II. Bacteriophages as cloning vectors

Bacteriophages are those viruses that are known to infect bacteria and structurally consist of only DNA (or RNA) surrounded by a protective protein coat called capsid. Basic mode of infection of phages generally begins with the attachment of phage outside the bacterium, followed by the injection of DNA into the host bacterium. Phage DNA molecule then replicates inside the host and capsid proteins are synthesized for assembly and release of new phages from the host (lytic cycle).

Depending upon the cycle followed by phages they are basically classified as:

- **Lytic Phages:** These phages are characterized by a rapid lytic cycle with entire life cycle of the phage being completed very quickly in nearly half an hour with lysis of the bacterial cell.
- **Lysogenic phages:** These are marked by a stable retention phase inside the bacterial cells. After the injection of phage DNA inside host cell it integrates with the host DNA (prophage) and is maintained in dormant state. This is then followed by lysis of bacterial cells and release of new phages in a similar manner to that of lytic phages.

Commonly used bacteriophage cloning vectors: M13 vector and Lambda (λ) vector

Before beginning with cloning a gene into the plasmid it is first required to understand a plasmid map and drawing a basic layout of cloning scheme. Here, we will restrict to the commonly used *E.coli* based plasmid vectors.

Nomenclature of plasmid vector: pBR322 "pBR322"

- **p** letter indicates that it is a plasmid.
- **BR** defines the laboratory where the vector was designed and constructed (BR represents Bolivar and Rodriguez, name of the two scientists who developed it).
- **322** is the number given to the plasmid to differentiate it from others designed in the same laboratory.

Characteristics of pBR322 as cloning vector:

- ORI: The origin of replication of pBR322 is originally from pMB1 (related to the colicin-producing plasmid ColE1)
- Size: Ideal size of 4363 bp, i.e., less than 10 kb as mentioned previously.
- Marker genes: It carries two selectable marker genes ampicillin (originally is from plasmid R1 of *E. coli*, and tetracycline (originally from R6-5 plasmid).

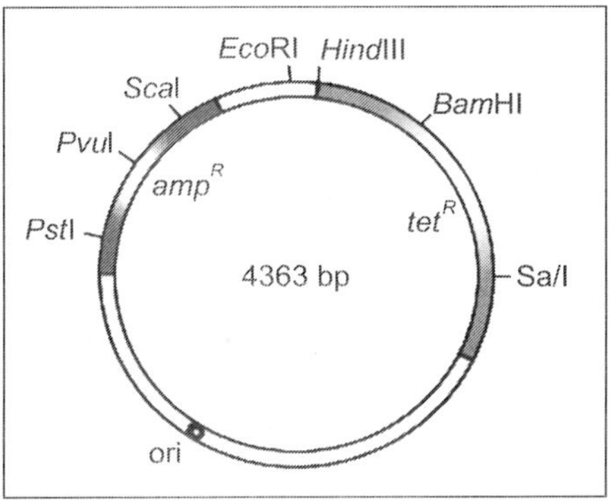

Fig. 15.1 Physical map of pBR322.
(*Source:* T.A. Brown)

- Cloning site: *Pst*I, *Pvu*I, or *Sca*I, *BamH*1 and *Hind*III mark few of restriction sites for manipulation.
- High copy number

Plasmid Map

A map showing the basic important features of the plasmid in a proper orientation is referred to as plasmid map. Following are a few points that are required, while drawing a plasmid map:

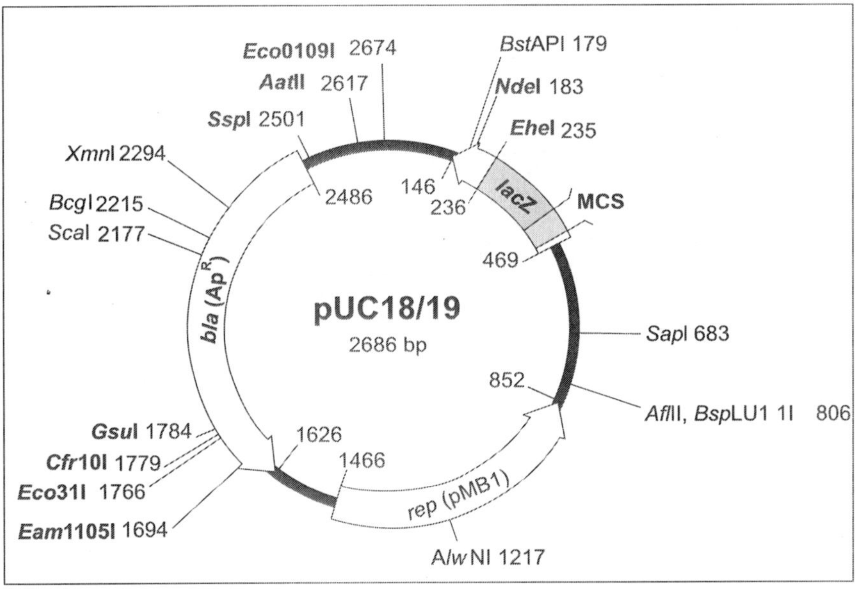

Fig. 15.2 Plasmid map of pUC18/19.
(*Source:* Fermentas Manual)

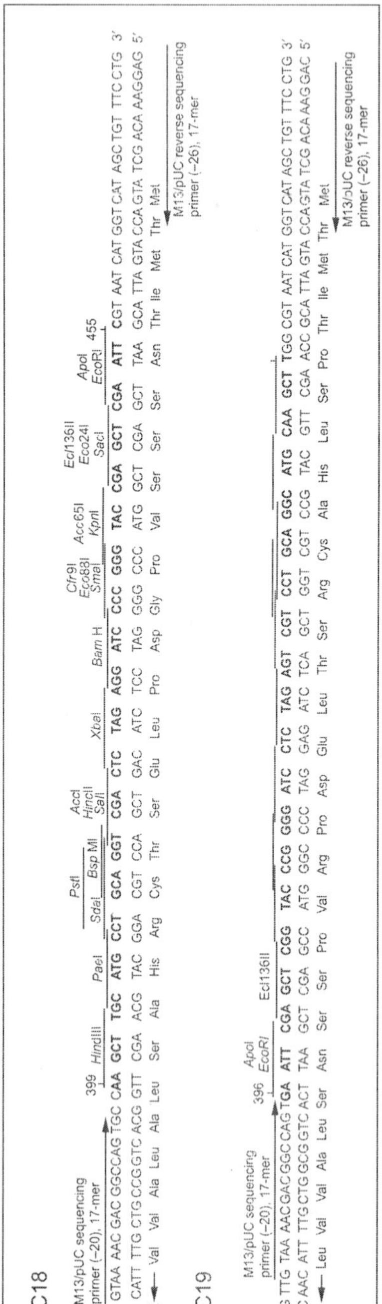

Fig. 15.3 MCS of pUC18/19.
(*Source:* Fermentas Manual)

- Name of the plasmid and its size are usually written in the centre of the plasmid circle.
- Only unique restriction sites are labeled on the plasmid map.
- Nucleotide positions of the restriction endonuclease sites are labeled next to the sites.
- Marker genes present are also written in correct orientation.
- Nowadays many software are also available that aid in drawing a plasmid map as this marks a key initial step for cloning.

General steps of cloning in plasmid vector

1. **Isolation of plasmid vector:** The very first step in gene cloning involves isolation of plasmid from *E. coli* (carrying amp resistance gene as well as lacZ gene coding for β-galactosidase, these properties are further implicated for selection of true amp resistant clones and blue-white screening) [Detailed protocol is mentioned in chapter on Plasmid Isolation.]
2. **Amplification of the desired insert DNA sequence:** The sequence of the gene of interest is PCR amplified using specific primers including the similar restriction sites as present on the MCS of the plasmid. [Detailed protocol for PCR amplification already described in the previous chapter.]
3. **Restriction digestion of the plasmid DNA, insert DNA and ligation:** The plasmid backbone and the amplified insert DNA are treated with the similar selected set of restriction enzymes. Restriction digestion of the plasmid results in generating a linearized plasmid from a circular one and treatment of the insert DNA with similar restriction enzymes leads to generation of ends that can be ligated together with that of plasmid DNA during ligation. [Detailed protocol for restriction digestion and ligation mentioned previously.]

   Note: During ligation there is a greater possibility of self-ligation of vector backbone as compared to the ligation of vector and insert DNA, thus in order to decrease this self-ligation digested vector is given phosphatase treatment that removes the terminal 5′-phosphate.
4. **Transfer of recombinant DNA (plasmid along with insert DNA) into the suitable host:** The ligated mixture of vector with insert DNA, also known as the recombinant DNA, is now transferred to the bacterial cells by the process of transformation. [Detailed protocol described in chapter on "Transformation".] Briefly, the bacterial cells are first made competent by neutralizing the

negative charge on membrane, followed by transformation of recombinant molecules during heat shock or electric current. With the division of transformed cells the recombinant DNA molecules are also passed on to the progeny cells, thus leading to "cloning".

5. **Screening of transformed cells:** In order to select only the transformed cells, the media for growing the transformed cells is amended with the appropriate antibiotic for the resistance marker allowing only resistant transformed cells to grow. However, such transformed cells may contain the correct recombinant DNA or self-ligated plasmid molecules. In order to distinguish between the true transformed cells with the recombinant DNA molecules from the false transformed cells possessing the self-ligated vector molecules the previously described system of blue-white screening or "lac selection" is implemented. [Described in detail in chapter on Transformation]. In addition to this positive clones can be identified by observing the DNA or protein expressed by the gene or the insert be identified by nucleic acid hybridization.
6. **Amplifying and purifying the recombinant DNA molecule:** This marks the final step of cloning; a positive colony screened above is aseptically transferred to the liquid medium and is allowed to grow under the antibiotic pressure. A fully grown culture is then harvested for isolation of the recombinant molecules. The recombinant molecules so obtained should be confirmed by restriction digestion and release of the insert to be observed by agarose gel electrophoresis or by sequencing of the recombinant molecule.

FREQUENTLY ASKED QUESTIONS

1. What is the relevance of unique restriction sites in the MCS of a plasmid vector?
2. What basic steps should be kept in mind while selecting the enzymes for restriction digestion during cloning?
3. How can the recombinant DNA molecule be confirmed for the correct insert?

REFERENCES

1. **Brown, T.** (2010). *Gene Cloning and DNA Analysis: An Introduction.* John Wiley & Sons.
2. **Sambrook, J., E.F. Fritsch, T. Maniatis** (1989). Enzymes used in Molecular cloning, in *Molecular Cloning: A Laboratory Manual.* 2nd Edn. (Editor: Nolan C) Cold Spring Harbor Laboratory Press.

3. **Yanisch-Perron, C., V. Jeffrey, M. Joachim** (1985). Improved M13 phage cloning vectors and host strains: nucleotide sequences of the M13mpl8 and pUC19 vectors. *Gene* **33:** 103–119.
4. http://dwb4.unl.edu/Chem/CHEM869N/CHEM869NLinks/www.fermentas.com/techinfo/NucleicAcids/mappuc1819.htm

16

DNA Blot Hybridization

Anjali Saxena & Udita Mukherjee
E-mail: *anjali.saxena1403@gmail.com*
udita297@gmail.com

INTRODUCTION

DNA blot hybridization or southern hybridization, one among the indispensable techniques till date in molecular biology, is widely used for detecting specific DNA sequences in DNA samples.

LEARNING OBJECTIVES

To be able to do the following

- Transfer DNA embedded in the agarose gel to oppositely charged nylon membrane
- Prehybridize the blot
- Prepare radioactive and non-radioactive labeled DNA probe
- Hybridize the blot with denatured radioactive or non-radioactive labeled probe
- Wash the blot
- Expose the blot to X-ray film and detection

REQUIREMENTS

Glassware and Plasticware

- Gel cutter
- Plastic wrap or parafilm
- One plastic container each, for wetting the membrane, setting up the blot and gel treatment, respectively
- Blunt-ended forceps
- Glass plates

Chemicals

- 0.25 M HCl solution for depurination

  Measured 490 mL of sterilized deionized water and added to it 10 mL of concentrated hydrochloric acid (HCl). The contents were poured into a flask that was sealed and subsequently swirled gently to mix the contents.

- 1.5 M NaCl, 0.5 M NaOH solution for denaturation
- 1 M TRIS, 1.5 M NaCl, pH 7.4, solution for neutralization.
- 20X SSC (3 M NaCl, 0.3 M Sodium citrate, pH 7.0)

Equipment

- Gel rocker
- Stop watch

Miscellaneous

- Positively charged nylon membrane
- Whatman filter paper (3 mm)

THEORETICAL ASPECT/PRINCIPLE

The southern blot hybridization technique still remains one of the most indispensable techniques of molecular biology. It mainly deals with the transfer of gel electrophoresed DNA to a membrane filter for detection of specific sequence by labeled complementary probes. It was the brain-child of Prof. E.M. Southern (1975). The basic principle involves fractionation of the DNA molecule according to the size on agarose gel and its subsequent electrophoretic transfer to nylon or nitrocellulose membrane. The DNA is fixed on the membrane by cross-linking using UV radiation and baking the membrane at 80°C. Radioactive or non-radioactive methods can be used to detect the signals from the membrane.

Hybridization involves the process of joining a single stranded DNA probe and the single stranded target DNA, now immobilized on the membrane, to form a double stranded DNA. The success of this experiment depends upon two factors mentioned henceforth.

1. the reactions are specific, i.e., the probes only bind to targets with complimentary sequence, and
2. the probe can recognise the specific target in a mixture of millions and binding to non-complimentary molecules does not occur.

General Controls to be Used in the Experiment

It is essential in every experiment to include positive and negative controls. Positive controls are essential in this case, in order to estimate the efficacy of hybridization. Negative controls are essential in order to identify the probable reasons for failure of the experiments and to rule out the possibility of contamination.

Positive control

The "correct" positive control/sensitivity control would be to dilute the probe, probably about a 1/1000 dilution based on the relative size × amount of the probe and the size × amount of the bacterial genome in order to get good signal.

Negative control

The negative control to be used is wild-type genomic DNA without the target gene, so that probe cannot find any complementary sequence and no signals should be obtained.

PROCEDURE

A. Transfer of gel electrophoresed DNA to nylon membrane

1. After visualizing the gel under UV, the blank areas of the gel were cut out with the gel cutter. For identification of the up and the downside of the gel, an appropriate cut was made on the top right hand corner of the agarose gel.
2. In one of the plastic containers, 250 mL of the **depurination** solution was poured and the gel was immersed into it. The gel was soaked for 15 min, exactly at room temperature with very gentle rocking on the gel rocker.
3. The depurination solution was poured off carefully, taking care that the gel didn't slip out of the container. It was then rinsed with appropriate amount of sterilized deionized water in order to remove the traces of HCl from the gel.
4. 250 mL of **denaturation** solution was poured gently into this container and the gel was soaked in it with gentle rocking for 15 min at room temperature.
5. The denaturation solution was poured off carefully and the above step was repeated. Once it was done, the gel was rinsed with appropriate amount of sterilized deionized water.
6. 250 mL of **neutralization** solution was poured into the same container and the gel was soaked in it with constant, gentle shaking for 15 min at room temperature.
7. The **neutralization** solution was poured off and the above step was repeated.
8. During the above incubation steps, the positively charged nylon was cut out of the same size as the size of the gel tray.

9. Three pieces of 3 mm Whatman paper were cut one cm larger than the size of the above nylon membrane. Another long piece of 3 mm Whatman paper was cut out of double the length of the above pieces, but having the same width (Fig. 16.1).

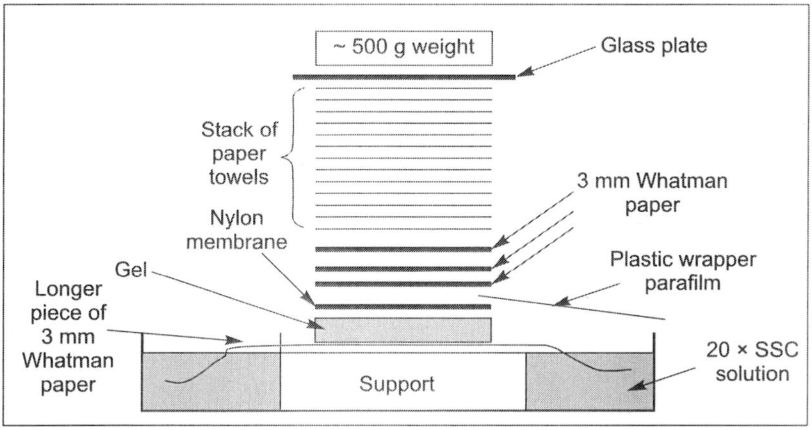

Fig. 16.1 Assembly for facilitating the capillary transfer of gel electrophoresed DNA to a nylon membrane.

10. The nylon membrane was completely wetted with sterilized deionized water first and then 20X SSC solution and was saturated in 20X SSC.
11. The transfer apparatus was assembled (see Fig. 16.1) for transferring DNA to the pre-wet nylon membrane.
    (a) The glass plate was kept over the support sitting in the container with 20X SSC solution.
    (b) The longer piece of 3 mm Whatman paper was wetted in 20X SSC solution and placed on the glass plate, avoiding the formation of bubbles. In case any bubbles were formed, they were rolled out by a pipette.
    (c) A little 20X SSC solution was poured at the center of the so far assembled apparatus.
    (d) The gel was placed **upside down** on the apparatus, carefully preventing the formation of bubbles.
    (e) The pre-wet nylon membrane was placed on top of the gel using blunt-end forceps.
    (f) Plastic wrap (or parafilm longer than the blot) were cut and placed on the four sides of the nylon membrane, and were stretched over the sides of the plastic container. Via capillary

transfer of 20X SSC, the DNA would be transferred from the gel to the nylon membrane.

(g) Three pieces of 3 mm Whatman paper were moistened with 20X SSC solution. These were then placed on top of the nylon membrane.

(h) A stack (~4-inches) of paper towels were placed on top of the apparatus assembled so far. The paper towels aid in absorbing water from the Whatman paper, resulting in upward pull of the water from the reservoir, facilitating **capillary** action.

(i) Finally, a glass plate was placed which was topped by a weight.

12. After leaving the assembly undisturbed overnight, the apparatus was gradually unassembled until the nylon membrane.
13. The impressions of the gel slots on the filters were marked.
14. The membrane was removed from the gel and was placed on a plastic wrap, cut out larger than the size of the blot.
15. The blot was subsequently exposed to UV radiation for 5 minutes so as to facilitate the cross-linking of the DNA to the nylon membrane.
16. The blot was then soaked in a 6X SSC solution for 5 min, with the side containing DNA facing upwards. The excess liquid from the membrane was soaked by placing it in between a wad of lint-free tissues.
17. The blot was baked at 80°C in a vacuum oven for 1.5 h, if it was not subjected to pre-hybridization treatments immediately. This was done to dry up the membrane as well as to fix DNA on it.
18. The blot was placed in a heat sealable bag right after it was baked.
19. The blot was stored until prehybridization was carried out.
20. If the blot was subjected to pre-hybridizing immediately right after step 16, they were transferred to a heat sealable bag.

B. Radioactive method of detection

The radioactive method involves labeling of nucleic acids by incorporating nucleotides containing radioisotopes (^{32}P, ^{33}P and ^{35S}) which emit β particles or γ rays. These rays can be detected by direct autoradiography which involves placing the sample in intimate contact with X-ray film. These β particles or γ rays emitted converts Ag+ ions to Ag atoms. The resulting latent image can be converted to visible image once the image is developed. Before dwelling on to dealing with radioactivity, follow appropriate safety protocols.

I. Pre-hybridizing the blots

- Dipped the membrane in 2X SSC and gently rolled the membrane with the help of forceps and placed in the roller hybridization bottle containing 50 mL pre-hybridization solution.
- Closed the bottle tightly and placed inside the hybridization oven at 65°C for 4–12 h.

II. Hybridization

Preparation of radiolabeled probe using Hexaprime labeling kit (Amersham Biosciences) in a microcentrifuge tube, added the following:

- Template DNA (20-50 ng) – x µL
- Random primers – 2.5 µL
- Milli-Q water - to make volume to 13 µL
- Denatured the contents at 100°C for 5 min in a boiling water bath.
- Kept the mixture at 37°C for 10 min for primer annealing.
- In another microcentrifuge tube added the following components:
- Reaction buffer : 2.5 µL
- dGTP : 2.0 µL
- dCTP : 2.0 µL
- dTTP : 2.0 µL
- Mixed both the above mixtures by pipetting and then added 2.5 µL of α-^{32}P dATP (BRIT, Hyderabad, India).
- Finally added 1 µL of Klenow enzyme and incubated at 37°C for 2 h.
- Stored the probe at – 20°C until its use.
- Added the probe to the hybridization bottle containing membrane in pre-hybridization solution and incubated at 65°C for 4–12 h.

III. Radioactive washing of the membrane:

- Removed the hybridization bottle containing the membrane from the hybridization oven and poured out the hybridization solution in a labeled falcon tube (to be reused) or discarded in a radioactive disposal sealed tank.
- Washed the membrane of excess radioactive solution by treating with 4 solutions one after the other at 65°C in the hybridization oven.
- 5X SSC/0.3% SDS : 10 min
- 2X SSC/0.1% SDS : 10 min
- 1X SSC/0.1% SDS : 15 min
- 0.1X SSC : rinse

- Removed the membrane from the hybridization bottle and placed on a tray containing 0.1X SSC.
- Removed the excess 0.1X SSC solution and carefully wrapped the membrane in a cling film.

Note: At this stage care should be taken to remove all the bubbles or it will create a background on X-ray film.

IV. Autoradiography

- The nylon membrane wrapped in cling film was placed in a tight cassette with intensifying screen.
- In the dark room, placed an X-ray film (Kodak) over the nylon membrane and closed the cassette.
- The cassette is kept at – 70°C for from 4–24 h depending on the half-life of the radioactivity intensifying the signals on the X-ray film.

Note: The activated crystals formed on the X-ray film due to exposure to beta rays are stabilized by keeping the cassette at – 70°C and thus increasing the sensitivity of the signals.

- After the required exposure, removed the X-ray film and dipped immediately in developer (Kodak) for 30 s to 1 min in the darkroom.
- Immediately transfer the film to fixer (Kodak) for 1 min.
- Finally washed the film in water and air-dried.

C. Non-radioactive method of detection

Radioactive tracers have the advantage of sensitive detection. However, they pose potential health hazards and must be handled carefully. Moreover radioactive waste is created, disposal of which is increasingly difficult and expensive. The non-radioactive methods of detection are thus gaining popularity. With the surge in demand, this method has undergone improvement and can compete with their non-radioactive counterparts in the field of sensitivity. This method involves detection either through chemiluminescence or chromogenic substrates which produce chemiluminescent signals or colored bands at the location of gene of interest.

DIG High Prime DNA Labeling and Detection Starter Kit I

This kit involves random primed DNA labeling with digoxigenin-dUTP and detection of hybrid by enzyme immunoassay (Roche Applied Science, Germany). The colored substrate reaction is brought about by combination of NBT (nitro-blue tertrazolium chloride) and BCIP (5-bromo-4-chloro-3′-indoylphosphate p-toluidine salt) to give a black purple precipitate when reacted with alkaline phosphatase, a popular enzyme conjugate for antibody probes. The method was carried out according to manufacturer's instructions.

I. Pre-hybridizing the blots

1. The membrane was rolled and placed in the hybridization bottles containing 10 ml of DIG Easy Hyb (provided with the kit- Table 1).
   (a) The amount of DIG Easy Hyb to be added is in accordance to the following calculation:
   (b) For 10 × 10 cm^2 of nylon membrane = 10 mL of DIG Easy Hyb
2. Closed the bottle tightly and placed inside the hybridization oven at 65°C for 1-3 h.

II. Hybridization

Preparation of probe:

3. 1 μg template DNA (the concentration of the DNA can vary from 0.6–1 μg) was taken.
4. The volume was made up to 16 μL with sterilized deionised water in sterilized microfuge tube.
5. The DNA was denatured by heating in a boiling water bath for 10 min and quickly chilled in ice.
6. Added 4 μL of DIG-High Prime (Table 16.1) to the denatured DNA mixed thoroughly by pipetting and giving it a flash spin in order to settle all the components at the bottom of the tube.

   Note: The vial was given a flash spin before using the DIG-High Prime.
7. The mix was incubated for 1 h or overnight.
8. The reaction was stopped by adding (200) of 0.2 M EDTA (pH 8.0). The probe was either stored at –20°C or used immediately.
9. The DIG labeled DNA probe (25 ng/mL DIG Easy Hyb) was denatured by boiling for 5 min and rapidly cooling in ice.

   Note: The probe is alkali labile and hence cannot be denatured by alkali treatment.
10. The denatured probe was added to pre-heated DIG Easy Hyb and mixed well. Care was taken to avoid foaming, as bubbles lead to background.
11. The membrane was kept in hybridization buffer for 16–18 h.

Note: The hybridization buffer can be reused. It is stored at – 20°C. Care was taken not to boil the hybridization buffer at any stage.

III. Non-radioactive washing of the membrane

12. The hybridization buffer was poured off.
13. The membrane was rinsed two times, for 5 minutes each, with 2X SSC/0.1% SDS at 15–25°C with constant agitation.

    Note: The first wash is low stringency wash to remove non-specifically bound probe.
14. The membrane was washed twice with 0.5X SSC/0.1% SDS at 65°C under continuous agitation. The duration of each wash was 15 min. The second wash is high stringency wash to remove undesired hybrids. Care was taken to pre-warm the buffer to correct temperature before adding to ensure that low homology hybrids are disrupted.

IV. Immunological detection

All the steps were performed at room temperature and care was taken not to let the membrane dry at any step.

15. After stringency washes the membrane was rinsed in washing buffer for 15–20 min.
16. The washing buffer was poured off and the membrane was incubated in 100 mL blocking solution (Table 16.1) for 2 h.

    Note: The blocking solution blocks unspecific binding sites and hence the long duration of incubation.
17. Incubated the membrane in 50 mL antibody solution (Table 1).
18. Washed the membrane twice in 100 mL washing buffer the duration of each was 15 min.
19. The membrane was equilibrated in 20 mL of detection buffer for about 5 min.
20. The detection buffer was poured off. The membrane was kept in a flat tray and 10 mL of freshly prepared color substrate solution was added onto the membrane. The reaction was carried out either in a dark container or was covered with aluminium foil.
21. The container with the membrane should be left undisturbed during the color substrate development.

    Note: The colored precipitate begins to form within a few minutes and usually completes in 16 h. The membrane can be exposed to light for short time interval to monitor the color development.
22. The reaction was stopped when the band of desired intensity was obtained by washing the membrane in 50 mL TE buffer for 5 min.

Table 16.1 Preparation of Pre-hybridization and Hybridization Solutions

Chemicals	*Composition*
1.5 M NaCl/0.5 M NaOH	87.6 g NaCl, 20 g NaOH per liter of deionized water
1 M Tris-HCl	121.14 g Tris per liter of deionized water. pH adjusted to 7.5 with conc. HCl
1 M Tris-HCl/ 1.5 M NaCl	121.14 g Tris; 87.6 g NaCl per liter of deionized water. pH adjusted to 7.5 with conc. HCl
20X Sodium salt of citrate (SSC)	3 M NaCl, 300 mM Tri sodium citrate (pH 7.0).
Pre-hybridization buffer	6X SSC, 0.5% SDS, 5x Denhardt's reagent, 100 mg/mL denatured fish sperm DNA (15 mL 20X SSC, 29 mL MQ, 2.5 mL 10% SDS, 2.5 mL 100X Denhardt's reagent, 1 mL denatured fish sperm DNA (5 mg/mL).
100X Denhardt's reagent	5 g ficoll, 5 g polyvinylpyrrolidone, 5 g bovine serum albumin-Sigma fraction V, 500 mL distilled water.
Salmon sperm DNA	Weigh the threads of Salmon sperm DNA for 100 mL, mix thoroughly in distilled water on a stirrer. Sonicate for 15–30 min till the solution becomes clear.
Antibody solution (DIG High Prime DNA labeling and Detection Starter kit I- Cat. No.11585614910)	Anti-Digoxigenin –AP conjugate (vial 4) (750U/mL) diluted 1:5000 times in Blocking solution (150 mU/mL)
Blocking solution (DIG High Prime DNA labeling and Detection Starter kit I- Cat. No.11585614910)	Diluted the content of (vial 6) in the ratio 1:10 in maleic acid buffer
Color substrate solution (DIG High Prime DNA labeling and Detection Starter kit I Cat. No. 11585614910)	200 μL in 10 mL Detection buffer
Detection buffer	0.1 (M) Tris-HCl pH 8 and 0.1 (M) NaCl; adjust the pH 9.5 with 1 (M) NaOH
DIG Easy Hyb working solution (DIG High Prime DNA labeling and Detection Starter kit I and II- Cat. No. 11585614910 and11745832910)	DIG Easy Hyb Granules (bottle 7) in 64 mL of deionized water.

Contd...

Contd...

Maleic Acid Buffer 0.1 (M)	Maleic acid, 0.15 (M) NaCl; pH 7.5
Washing buffer	0.1 (M) Maleic acid, 0.15 (M) NaCl and 0.3% Tween 20
Washing Solution-I (non-radioactive detection)	SSC (2X) and SDS (0.1%) in sterile deionized water
Washing Solution-II (non-radioactive detection)	SSC (0.5X) and SDS (0.1%) in sterile deionized water

RESULTS

An example of a southern blot is shown in Fig. 16.2.

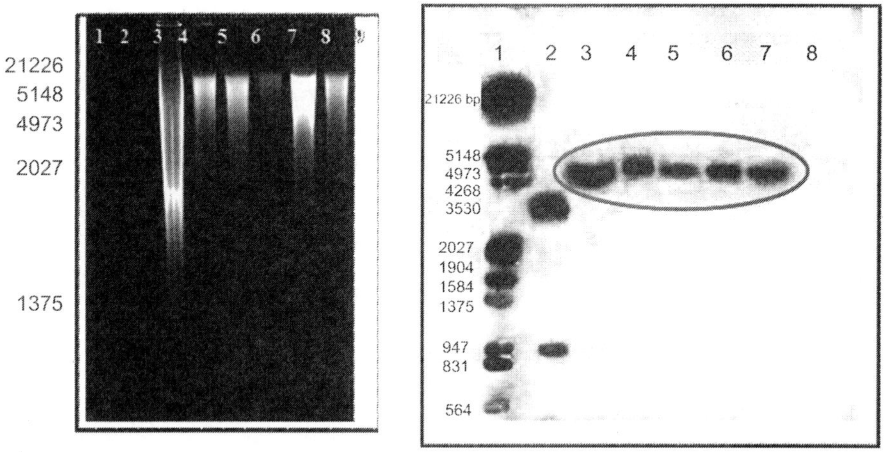

Fig. 16.2 An example showing southern blot: (A) 0.8% agarose gel electrophoresis gel showing restriction profile of genomic DNA and (B) the corresponding southern blot hybridization result. Lane 1: lambda DNA ***Eco*****RI/*****Hind*****III** digested (bp) (MBI Fermentas #SMO191), Lane 2: gDNA (digested) as positive control, Lane 3-7: gDNA from experimental samples, Lane 8: gDNA as negative control. Positive signals obtained are encircled in red. No signal is seen in negative control.

DISCUSSION

As evident from the gel picture Fig. 16.2, the amount of DNA (concentration) should be good so as to have good signals on the blot. A blot is a replica of gel that shows the gene of interest in the form of black spots as positive signals. Positive and negative controls are important as they help to figure out the authenticity of experiment.

FREQUENTLY ASKED QUESTIONS

1. How can you label probe for southern blot analysis?
2. What is southern dot blot analysis and what is it used for?
3. What two hybridization parameters or conditions are typically altered for maximum hybridization?
4. What kinds of nucleic acid bind to nitrocellulose? What kinds of nucleic acid do not bind?
5. Why do you soak the gel and not the colony lift in 0.2 N HCl?
6. Explain how the DIG-labeling and detection kit that you used works.

REFERENCE

1. **Southern, E.M.** (1975). Detection of specific sequences among DNA fragments separated by gel electrophoresis. *J. Mol. Biol.* **98:** 503.

17

DNA Sequencing

Anukriti Sharma, Utkarsh Sood & Mallikarjun Shakarad

E-mail: *anukriti.sharma8@gmail.com*

utkarsh.sood@gmail.com

beelab.ms@gail.com

INTRODUCTION

It is rare for any year to go by without some new version of DNA sequencing technology coming up. Until today, many genomes have been sequenced ranging from bacteria to humans. Newer sequencing technologies can vary, although, the basics being extraordinarily simple. DNA structure comprises a long chain of alternating sugars and phosphates, with each sugar connected to one of the four bases. The technique by which the order of the nucleotide bases in DNA (adenine, guanine, cytosine, and thymine) is determined using various sequencing methods is called DNA sequencing.

As simple is the structure, simpler is to understand the fundamentals of sequencing. Complexity comes when we want to address high throughput sequencing, i.e., sequencing a lot of DNA molecules at the same time. High throughput sequencing is dealt with next-generation sequencing (NGS).

LEARNING OBJECTIVES

- To understand the process of dideoxy DNA sequencing (Sanger sequencing)
- To understand complementary base pairing and apply it to DNA sequencing
- To understand the components required for dideoxy DNA sequencing
- To understand the basics of polymerase chain reaction (PCR) and its use in DNA sequencing
- To revisit the components and molecular reactions occurring in chain termination sequencing

COPYING OF DNA

DNA structure is made up of three components: a backbone consisting of **s**ugars and **p**hosphates. Sugar is linked to one of the four bases: adenine

(A), guanine (G), cytosine (C), and thymine (T) out of which adenine and guanine fall into purine group and cytosine and thymine into pyrimidine group (Fig. 17.1). DNA has double stranded alpha helical structure characterized by hydrogen bonding between a purine and pyrimidine, i.e., A pairs with T by two–H bonds, and G pairs with C by three–H bonds. The strategy of genome sequencing is described in Fig. 17.2.

The replication of DNA, i.e., amplifying a single or few pieces of DNA across several orders of magnitude, leading to generation of thousands to millions of copies of DNA, which can be further used for sequencing are generated using a molecular biology technique called polymerase chain reaction (PCR) developed by Kary Mullis in 1983.

PCR contains seven essential components:

- A thermostable DNA polymerase which can perform the polymerization of the template DNA, e.g., Taq polymerase.
- A pair of primers (synthetic oligonucleotide) to initialize DNA synthesis (forward and reverse primers).
- Deoxynucleoside triphosphates (dNTPs) as they are the building units of DNA.
- Divalent cations like Mg^{2+} as cofactor.
- Buffer to maintain pH as buffers neutralize the acids and bases in their buffering range.
- Monovalent cations like K^{+} which stabilizes primer annealing.
- Template DNA, i.e., DNA which is to be sequenced

PCR is an iterative process, consisting of the following steps:

1. **Denaturation:** It is the first step in systematic polymerase chain reaction event in which the reaction mix is heated to 96°C for 45 s. DNA template undergoes denaturation (DNA melting) due to the disruption of hydrogen bonds between complementary bases of the double stranded DNA, generating single-stranded DNA molecules.
2. **Annealing of primers to the template DNA:** In the second step, the temperature of reaction mix is lowered to 50–65°C for 20–45 s to allow the primers to anneal to the single-stranded DNA template. The annealing temperature and the time duration of this step depends on the sequence of the primer and their respective templates. The annealing temperature is about 3–5°C below the meting temperature (Tm) of the primers used. When the primer sequence matches closely with the template sequence, the primer gets annealed to its respective DNA template by the formation of stable H-bonds between them. The primer-template hybrid is identified by the polymerase which then amplifies the template DNA.

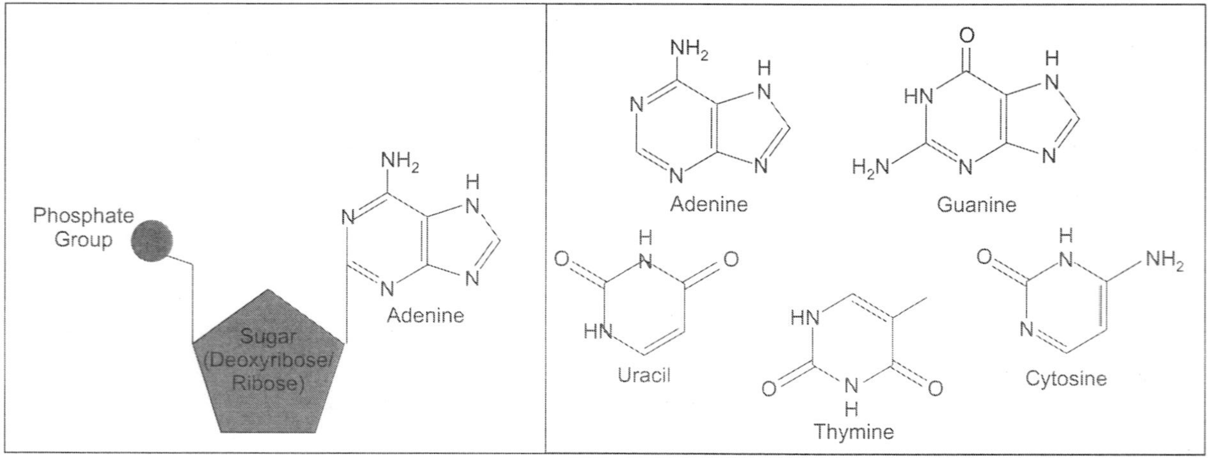

Fig. 17.1 Nucleotide of DNA (left). The nitrogenous bases found in nucleic acids (right).

3. **Extension of oligonucleotide primers:** During this step, a new DNA strand complementary to the DNA template strand is synthesized by DNA polymerase. The temperature at this step is kept at 72°C which is the most optimum for the activity Taq DNA polymerase. The polymerase adds dNTPs that are complementary to the template in 5′ to 3′ direction, by forming a phosphodiester bond between the 5′-phosphate group of the dNTPs with the 3′-hydroxyl groups at the end of the extending DNA strand. The duration of this step depends on the DNA polymerase used and on the length of the DNA fragment to be amplified. Generally, 30 s are required to amplify a segment of 500 kb. The amplification proceeds exponentially, doubling the amount of target DNA at each cycle unless substrates or reagents are limited.
4. **Final elongation:** At this step, the reaction mix is maintained at a temperature of 72°C for 7 min after the last PCR cycle to confirm that the entire single-stranded DNA is fully extended.
5. **Number of cycles:** The number of cycles which are needed to carried out depends on the number of copies of template DNA

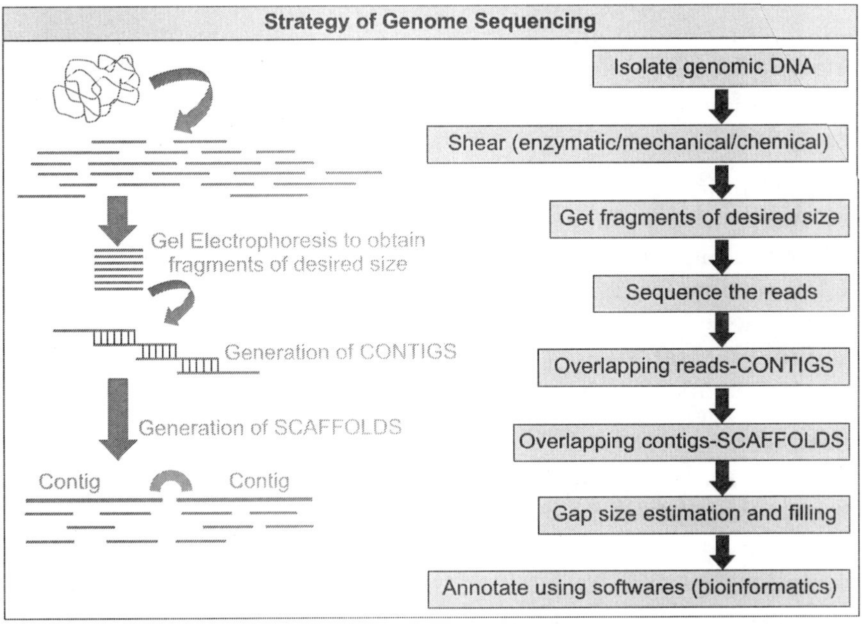

Fig. 17.2 Flow chart representing the strategy of genome sequencing.

present in the beginning of the reaction, also the efficiency of primer extension and amplification. To achieve acceptable levels of amplification of single copy target DNA sequences, at least 25 cycles are required, but one should always assure that 30 cycles are carried, so the product can be easily eluted from the gel.

Sequencing techniques include:

1. Maxam Gilbert sequencing (chemical sequencing)
2. Sanger sequencing (dideoxy chain termination sequencing)
3. Pyrosequencing
4. GS FLX 454
5. Illumina sequencing
6. SoLiD™ system sequencing

Allan Maxam and Walter Gilbert in 1977 developed a method for sequencing DNA molecules based on chemical modification of DNA and subsequent cleavage of the DNA backbone at sites adjacent to the modified nucleotides. This method is known as chemical sequencing method or simply Maxam Gilbert sequencing method on the name of its inventors.

Maxam–Gilbert sequencing and Sanger's chain termination method were the first widely-used methods for DNA sequencing and represent the first generation of DNA sequencing methods.

The strategy used in Maxam-Gilbert sequencing is given in the following flow charts (Fig. 17.3 and Fig. 17.4).

Polyacrylamide Gels: The ssDNA fragments are visualized using high resolution polyacrylamide sequencing gels. The ssDNA fragments are fractionated on the basis of their size and a high resolution sequencing gel can resolve fragments which differ by a single base in length. The gel routinely contains 6–20% polyacrylamide and 7 M urea which denatures the secondary structures of DNA and help to maintain the ssDNA to proper size fractionation.

Drawbacks of Maxam-Gilbert Sequencing

- High quantity of purified DNA is required, and the method contains many intermediate purification steps.
- The read lengths are relatively short, around 100 bp from labelled ends.
- Automation not available (sequencers).
- It takes lot of time to produce the sequence as it is first end-labelled and re-isolated prior to the actual sequencing.

Therefore, Maxam–Gilbert sequencing is no longer used by researchers, and has been replaced by new next-generation sequencing methods.

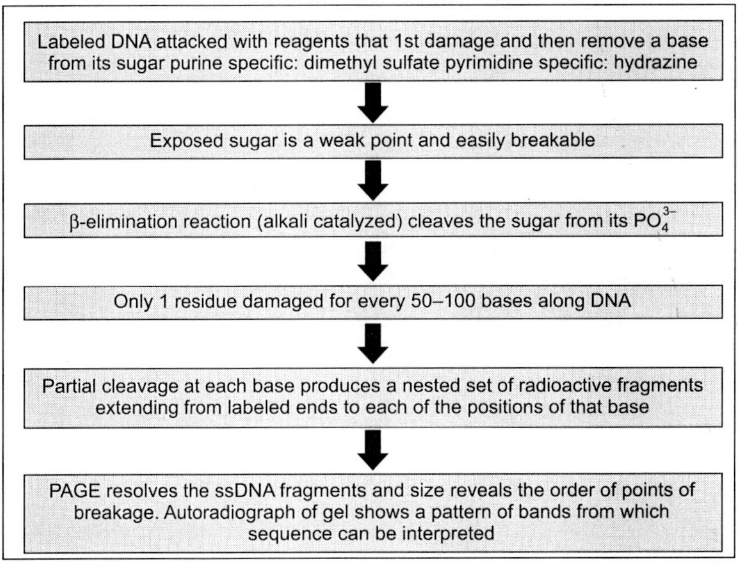

Fig. 17.3 Flow chart representing the strategy used in Maxam-Gilbert sequencing.

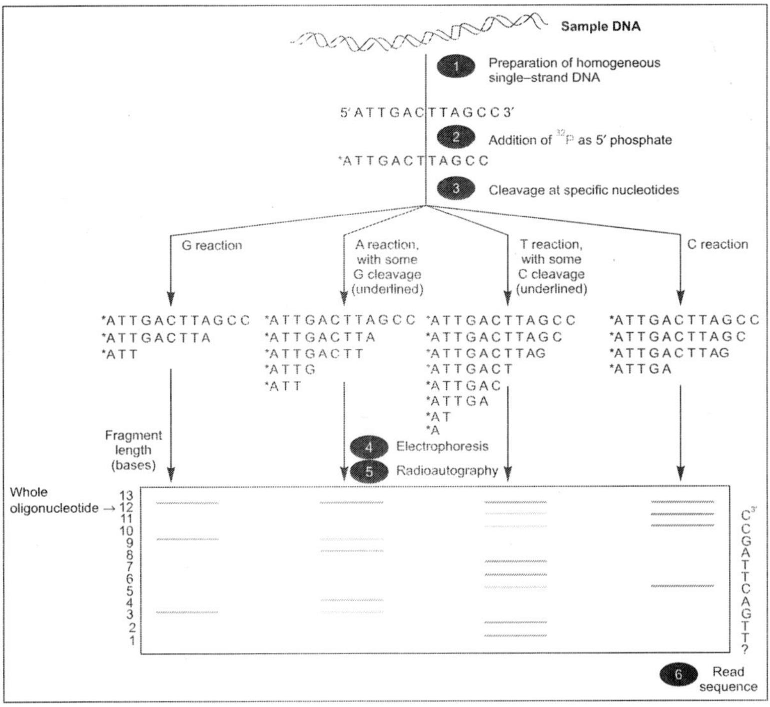

Fig. 17.4 Overview of the entire protocol performed during Maxam-Gilbert sequencing.

Source: Mathews and van Holde: *Biochemistry* 2/e.

High throughput sequencing

Sanger Sequencing

Frederick Sanger and his colleagues in 1977 developed a method for DNA sequencing based on chain terminators, i.e., dideoxynucleotides (ddNTP). These nucleotides lack a 3′-OH group which is required for formation of a phosphodiester bond between two nucleotides causing DNA polymerase to stop extending upon addition of a ddNTP.

DNA sample is divided into 4 reactions containing DNA polymerase and all dNTPs (dATP, dGTP, dCTP, dTTP). In each reaction, one ddNTP is added with rest 3 ordinary dNTPs. This is then followed by standard procedure of PCR, and resulting fragments are run on gel electrophoresis in 4 separate lanes on gel. Chain termination as a result of incorporation of a dideoxynucleotide (ddATP, ddGTP, ddCTP, or ddTTP) in a DNA fragment can be seen as a band in a lane. Therefore, relative positions of these bands along the four lanes can be used to read the DNA sequence from bottom to top.

Variations to Sanger methodology

Technical variations of the conventional method of chain termination developed by Leroy Hood and coworkers (1986), in terms of fluorescently

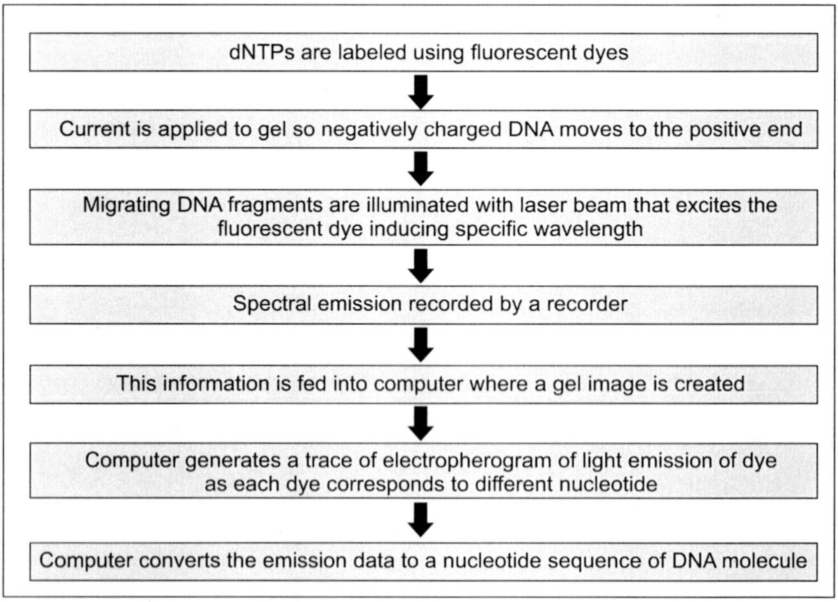

Fig. 17.5 Flow chart representing the protocol for automated DNA sequencing using fluorescent dyes.

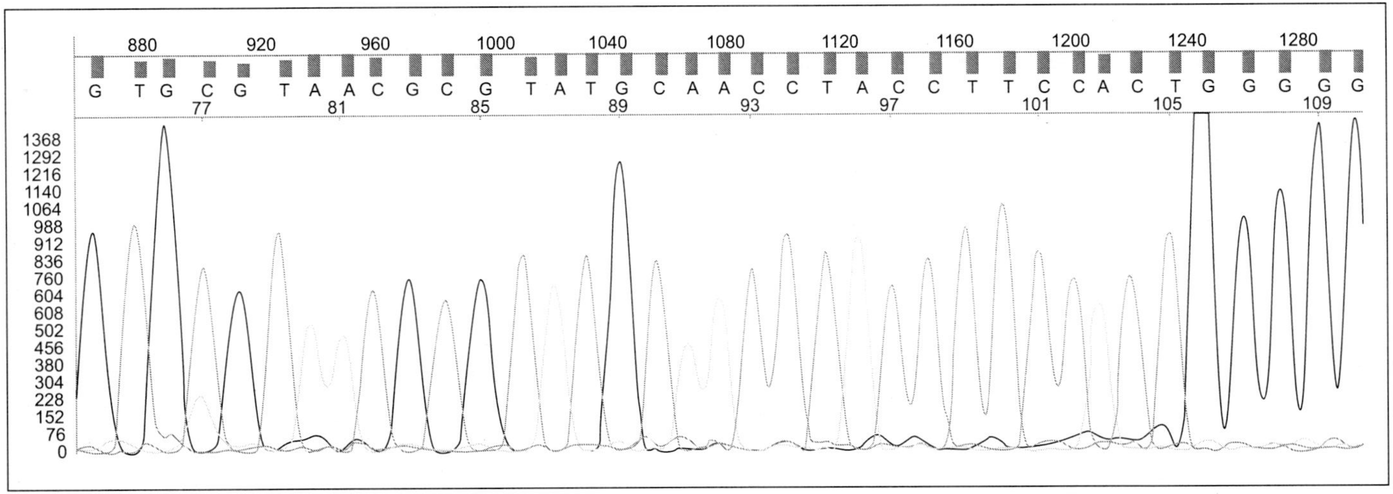

Fig. 17.6 Chromatogram generated after Sanger sequencing by automated base calling software. Each base is color coded with different colors.

labeled ddNTPs further facilitated the reading of nucleotides in single reaction for faster automation (Fig. 17.5).

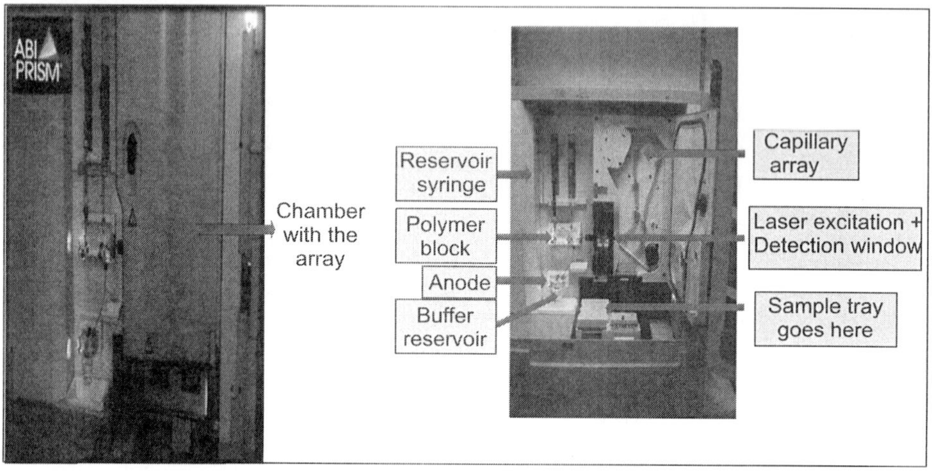

Fig. 17.7 3100 Avant genetic analyzer at Central Instrument Facility at Department of Zoology, University of Delhi.

PROCEDURE

Cycle sequencing reaction

The DNA required for cycle sequencing is ~150–200 ng. 1µL aliquots from the precipitated DNA was checked on 0.8% agarose gel loaded with 100 ng of λ-*Hind*III marker for visual quantification.

ABI PRISM BigDye® Terminator v3.1 cycle sequencing kit (Applied Biosystems, USA) was used for performing the sequencing reaction. This kit includes BigDye® Terminator v1.1/3.1 (containing AmpliTaq FS DNA polymerase, dNTPs, fluorescently labelled ddNTPs, magnesium chloride) and sequencing buffer (5X).

For setting up the reaction, an appropriate volume of DNA was aliquot in MicroAmp 96-well reaction plate (Applied Biosystems, USA) using multichannel pipette. A master mix was prepared containing the BigDye® Terminator v3.1, sequencing buffer, primer N/ primer Sand sterile Milli-Q water. This master mix was transferred to the reaction plate to make up the volume to 10 µL per well.

Cycle Sequencing PCR

Reaction for one well of 96-well plate was set up as follows:

Template DNA - 2.0 µL

BigDye® Terminator v3.1 - 0.5 µL

BigDye Sequencing Buffer (5X) - 2.0 μL
*Primer (100 uM) - 0.06 μL
Sterile Milli-Q water - 5.5 μL
Total - 10.0 μL

(*Primer N and primer S were added in two separate reactions.)

For this chapter, each precipitated DNA plate was sequenced from both the ends using Tn7 L (or Primer S) and Tn7 R (or Primer N) primers. The reader can design the primer for his own need by using the method in Chapter 9. The sequences of the primers used in this chapter are as follows:

Primer N: 5′ -ACTTTATTGTCATAGTTTAGATCTATTTTG- 3′
Primer S: 3′ -TCCCCACCTTTACCTCAAAAATTCCTAATA- 5′

The samples were spun down by brief centrifugation to bring down the components and the cycle sequencing reaction was set up using the following conditions:

Denaturation, 95°C - 10 s
Annealing, 55°C - 5 s
Extension, 62°C - 45 s
Final Extension, 72°C - 7 min
Cycles - 30

Clean-up of the Extension Products

Before loading the samples in the sequencer, the incorporated dye terminators are completely removed using clean-up protocol.

1. 10 μL of Milli-Q water and 5 μL of 125 mM EDTA (pH = 8.0) was added in each well of 96-well microtitre plate containing cycle sequencing product.
2. 85 μL of ethanol was added to each well and the plates were sealed with scotch pads. Plates were invert mixed 2–3 times.
3. Plates were incubated at room temperature for 15 min and centrifuged at 3500x g for 30 min at 18°C.
4. The supernatant was decanted by gently inverting the plates on a tissue paper towel and removed the remaining precipitation solution by giving the plates a brief spin at 18x g in an inverted position over a tissue paper towel.
5. 150 μL of 70% ethanol was added to each well and centrifuged at 3500x g for 15 min at 18°C.
6. The supernatant was removed as described previously, following which the samples were air-dried for 15 min and finally dissolved in 10 μL of Hi-Di™ Formamide (Applied Biosystems, USA).

7. The plates were incubated at 37°C for 15 min with gently shaking the contents that were brought to the bottom of the well by brief spinning.

Linking Plate to the ABI Sequencer (Fig. 17.7)

1. The black plate holder was kept on the platform of instrument.
2. The plate was locked. (If locking was correct, then a green light will blink.)
3. 'Plate import files' option was opened from the desktop of associated computer terminal.
4. The respective folder where sequencing results shall be saved was opened.
5. Any existing parameter file was opened.
6. The plate was renamed and 'Saved As' a new file.
7. 'Import' option was clicked to check all the parameters stored in the working file as:
   - Sample No.: N
   - Sequencer Dye used: H (for Hi-Di™ Formamide)
   - Mobility file: DT 3700 POP6 (BDv3) v1.mob (for model of instrument used and polymer)
   - Project name: 3700 Project N
   - Run Module 1: Seq2_POP 6 Default Module
   - Analysis Module: BC_3700 POP 6_Seq

Base Calling of Sequence Reads

After completion of the sequencing run, the sequence quality and the average read length is analyzed using Trace Tuner software (Paracel® Ic., USA). It is used to review the ABI electropherograms (in Fig. 17.6) and base calls. It also adjusts these base calls and assigns quality values to each. The quality values correspond to predictions of the probability of each base being called in error, based on certain characteristics of the associated electropherogram peak and how it relates to the peaks around it. Trace Tuner software uses the raw chromatogram data, which is in the form of *.ab1 files. The definitions of quality values are those used by Phred, hence, it can similarly be used in the first step in a bioinformatics pipeline as:

1. The Trace Tuner software was opened from its location in terminal.
2. Moved to the folder containing the ABI electropherograms {e.g./AppliedBio/3700/*DataExtractor*/sequencing run (e.g., Run_demo 3700_2009-10-03_3)}.

3. All files with extensions (*.ab1) were selected and 'OK' was clicked.
4. 'Run' was clicked.
5. Window will display message as 'assigning quality scores'.
6. Finally, an average score is provided.

FREQUENTLY ASKED QUESTIONS

1. What kind of templates can be sequenced?
2. Can we sequence genome in a single stretch?
3. Why polyacrylamide gels should be used instead of agarose gels for sequencing?
4. What are the molecular components used to set up a PCR reaction?

REFERENCES

1. **Maxam, A., W. Gilbert** (1977). A new method for sequencing DNA. *Proc. Natl. Acad. Sci.* **74:** 560–564.
2. **Ronaghi, M.** (2001). Pyrosequencing sheds light on DNA sequencing. *Genome Res.* **11:** 3-11.
3. **Sanger, F., S.Nicklen, A.R.Coulson** (1977). DNA Sequencing with chain-terminating inhibitors. *Proc. Natl. Acad. Sci.* **94:** 5463–5467.
4. **Shendure, J., H. Ji.** (2008). Next-generation DNA sequencing. *Nature Biotech.* **26:** 1135-1145.
5. **Venter, C., M. D. Adams, E. W. Myers, P. W. Li, R. J. Mural *et al.*** (2001). The sequence of the human genome. Science **291:** 1304-1351.